Intelligent Systems Reference Library

Volume 112

Series editors

Janusz Kacprzyk, Polish Academy of Sciences, Warsaw, Poland
e-mail: kacprzyk@ibspan.waw.pl

Lakhmi C. Jain, University of Canberra, Canberra, Australia;
Bournemouth University, Poole, UK;
KES International, UK
e-mails: jainlc2002@yahoo.co.uk; Lakhmi.Jain@canberra.edu.au

About this Series

The aim of this series is to publish a Reference Library, including novel advances and developments in all aspects of Intelligent Systems in an easily accessible and well structured form. The series includes reference works, handbooks, compendia, textbooks, well-structured monographs, dictionaries, and encyclopedias. It contains well integrated knowledge and current information in the field of Intelligent Systems. The series covers the theory, applications, and design methods of Intelligent Systems. Virtually all disciplines such as engineering, computer science, avionics, business, e-commerce, environment, healthcare, physics and life science are included.

More information about this series at http://www.springer.com/series/8578

Aleksandra Klašnja-Milićević
Boban Vesin · Mirjana Ivanović
Zoran Budimac · Lakhmi C. Jain

E-Learning Systems

Intelligent Techniques for Personalization

 Springer

Aleksandra Klašnja-Milićević
Faculty of Sciences, Department of Mathematics
  and Informatics
University of Novi Sad
Novi Sad
Serbia

Boban Vesin
Department of Computer Science and Engineering
University of Gothenburg/Chalmers, University of
  Technology
Gothenburg
Sweden

Mirjana Ivanović
Faculty of Sciences, Department of Mathematics
  and Informatics
University of Novi Sad
Novi Sad
Serbia

Zoran Budimac
Faculty of Sciences, Department of Mathematics and
  Informatics
University of Novi Sad
Novi Sad
Serbia

Lakhmi C. Jain
University of Canberra
Canberra
Australia

and

Bournemouth University
Poole
UK

and

KES International
UK

ISSN 1868-4394   ISSN 1868-4408   (electronic)
Intelligent Systems Reference Library
ISBN 978-3-319-82284-6   ISBN 978-3-319-41163-7   (eBook)
DOI 10.1007/978-3-319-41163-7

This Springer imprint is published by Springer Nature
The registered company is Springer International Publishing AG Switzerland

Foreword

In past decade a lot of efforts have been put into development of e-learning: many systems and repositories of learning objects have been developed, a notion of learning object has been introduced and spread, learning object metadata standards have been released, interoperability of e-learning system components have been discussed. Thousands of papers, thesis and various research works on e-learning are published every day. It seems that a new journal on e-learning opens every week. Universities already embrace the power of e-learning to deliver content to students all over the world, even for free.

E-learning, as an important segment of educational environments, represents a unique opportunity to learn independently, regardless of time and place, to acquire knowledge without interruption and based on the principles of traditional education. E-learning offers a number of advantages for people and companies looking to develop a new content programme or curricula. That is, whether your trainees are all together in a classroom or scattered all over the country in different time zones, they can still tap into the same course materials, and at a time that's convenient to them.

One of the most important segments in today's development and use of the e-learning system is the personalization of content and building of user profiles based on the learning behaviour of each individual user. The personalization options increase efficiency of e-learning, thus gaining much acceptance as it allows the learners to set their own goals, learn at their own pace, and even decide on their method of learning thus leading to better learning results. In order to personalize the learning process and adapt content to each learner, e-learning systems can use different strategies that have the ability to meet the needs of learners.

This monograph provides a comprehensive research review of intelligent techniques based on the modern perspective of research and innovation for personalization of e-learning systems. Personalized learning approach promotes a tailored support system, helping learner to learn. In order to personalize learning, one needs to personalize learning objects and their modules and courses, learning activities and learning environments.

Special emphasis is given to intelligent tutoring systems as a particular class of e-learning systems, which support and improve the learning and teaching of domain-specific knowledge. Each of the subsequent chapters of this monograph reveals leading-edge research and innovative solution that employ personalization techniques with an application perspective.

It is obvious that different learners have different preferences, needs and approaches to learning. Psychologists distinguish these differences as individual learning styles. Learning styles can be defined as unique manners in which learners begin to concentrate on, process, absorb and retain new and difficult information. Therefore, it is very important to accommodate for the different styles of learners through learning environments that they prefer and find more efficient. Furthermore, in modern Web-based learning environments, the authors avoid creation of static learning material that is presented to the learner in a linear way, due to the large amount of interdependences and conditional links between the various pages. Often, authors create multiple versions of learning resources so the system can propose to the learner the appropriate one. This leads to the learning concept known as content adaptation.

An important part of the personalization possibilities is certainly the prospect of using the recommender system. Recommender system can be defined as a platform for providing recommendations to users based on a specific type of information filtering technique that attempt to recommend information items (movies, music, books, news, Web pages, learning objects and so on). Recommender systems strongly depend on the context or domain they operate in, and it is often not possible to take a recommendation strategy from one context and transfer it to another context or domain. Personalized recommendation can help learners to overcome the information overload problem, by recommending learning resources according to learners' habits and level of knowledge. The first challenge for designing a recommender component for e-learning systems is to define the learners and the purpose of the specific context or domain in a proper way.

To improve recommendation quality, metadata such as content information of items has typically been used as additional knowledge. With the increasing popularity of the collaborative tagging systems, tags could be interesting and useful information to enhance algorithms for recommender systems. Collaborative tagging systems have grown in popularity over the Web in the last years based on their simplicity to categorize and retrieve content using open-ended tags.

A recent trend in the field of e-learning and tutoring systems is to utilize agent technology, develop and use different kinds of agents with various degrees of intelligence, capable of exhibiting both reactive and pro-active behaviour in order to satisfy their design goals in virtual learning environments. The monograph presents a possible trend in use of intelligent agents for personalized learning within tutoring system.

The validity of viewing e-learning initiatives' development from an information systems' perspective is supported by recognizing that both of these efforts are fuelled by a common goal to harness new technologies to better meet the needs of their users.

I believe that the chapters presented in the monograph update on the modern perspective of the education environments and personalization techniques per research and innovation, and are beneficial for designing better e-learning systems. I have recognized the significance of the monograph for researcher, practitioners and students interested in the personalized e-learning technology. I expect it will motivate and encourage new issues and challenges for the future scientific research in this field.

This book is fascinated read for students of all levels and teachers, also for those curious to learn about the e-learning in a systematic way.

Prof. Valentina Dagienė
Vilnius University, Lithuania

Preface

The rapid development of the contemporary new Web technologies and methods made online education increasingly accessible, open and adaptable; allowed new techniques, approaches and models to emerge and reasoned the revolution in the digital knowledge age that enabled greater and faster human (social) communication and collaboration and led to fundamentally new forms of economic activity that produced the knowledge economy and required changes in education. The increasing need for quality education requires expertise which is continually being developed. The integration of e-learning (short form of Electronic Learning) into the education system is viewed as one way to meet this growing need for high-quality education.

This monograph brings a result of our attempts to represent the most important aspects of current theory and practice in emergent e-learning approaches, systems and environments. As a specific case study we will present in details Web-based tutoring system we have been developing for last several years. This system incorporates a lot of contemporary techniques and methods from e-learning and technology-enhanced learning areas.

The material covered in the monograph is addressed to students, teachers, researchers and practitioners in the areas of e-learning, recommender systems (RS), semantic Web and machine learning.

This monograph is organized into five major parts. Part I: *Preliminaries*, which includes Chap. 1 of the monograph—*Introduction*, introduces the motivation and objectives studied in the subsequently presented research, and presents major standards and specifications in e-learning.

Part II: *E-learning Systems Personalization*, which consists of Chaps. 2–7, provides an overview of personalization techniques in e-learning systems. Chapter 2—*Personalisation and Adaptation in E-Learning Systems* shows the most popular adaptation forms of educational materials to learners. Chapter 3—*Personalisation Based on Learning Styles* presents the bases of electronic learning techniques for personalization of learning process based on individual learning styles and the possibilities of their integration into e-learning systems. The most popular adaptation techniques used in e-learning environments are presented in Chap. 4. Following chapter—*Agents in E-Learning Environments*—presents current trends in use of intelligent agents for personalization. Chapter 6—*Recommender Systems in E-Learning Environments*—provides an overview of techniques for recommender systems, folksonomy and tag-based recommender systems to assist the reader in understanding the material which follows. The overview, presented in Chap. 7 includes descriptions of content-based recommender systems, collaborative filtering systems, hybrid approach, memory-based and model-based algorithms, features of collaborative tagging that are generally attributed to their success and popularity, as well as a model for tagging activities and tag-based recommender systems.

Part III: *Semantic Web Technologies in E-Learning* contains a review of the basic elements of semantic Web, as well as the possibilities of applying semantic Web technologies in e-learning. Chapter 9—*Design and Implementation of General Tutoring System Model*—displays the details of a general tutoring system model, supported with semantic Web technologies as well as the principles of creating courses in different domains supported by this model.

Part IV: *Case Study: Design and Implementation of Tutoring System*, which consists of Chaps. 10 and 11, presents the most important requests for implementation of personalization options in e-learning environments, as well as design, architecture and interface of Protus 2.1 system. Chapter 10 presents the details about previous versions of the system, defined user requirements for the new version of the system, architecture details, as well as general principles for application of defined general tutoring model for implementation of programming courses in Protus 2.1. Chapter 11 presents Protus 2.1 functionalities as well as personalization options from the end-user perspective.

Part V: *Evaluation and Discussion*, which contains Chaps. 12 and 13, highlights the results of the evaluation and discussion of analysis of the results regarding the validity of the system. Finally, Chap. 13 concludes this monograph, summarizing the main contributions and discussing the possibilities for future work.

Contents

Part I Preliminaries

1 Introduction to E-Learning Systems   3
   1.1   Web-Based Learning   4
   1.2   E-Learning   4
   1.3   E-Learning Objects, Standards and Specifications.   7
        1.3.1   E-Learning Objects.   8
        1.3.2   E-Learning Specifications and Standards   11
        1.3.3   Analysis of Standards and Specifications.   15
   References   16

Part II E-Learning Systems Personalization

2 Personalization and Adaptation in E-Learning Systems   21
   2.1   Personalization and Personalized Learning   22
   2.2   Adaptation of E-Learning Systems   23
   References   25

3 Personalization Based on Learning Styles   27
   3.1   Learning Style's Theories   28
   3.2   Learning Styles in E-Learning Systems.   31
   3.3   Learning Style Index by Felder and Soloman   33
        3.3.1   Information Processing: Active and Reflective
              Learners   34
        3.3.2   Information Perception: Sensing and Intuitive
              Learners   34
        3.3.3   Information Reception: Visual and Verbal Learners . . .   35
        3.3.4   Information Understanding: Sequential and Global
              Learners   35
   References   35

4   Adaptation in E-Learning Environments.   37
      4.1   Adaptive Educational Hypermedia   38
      4.2   Content Adaptation   39
      4.3   Link Adaptation.   40
      References   42

5   Agents in E-Learning Environments.   43
      5.1   Some Existing Agent Based Systems   44
      5.2   HAPA System Overview   45
             5.2.1   Harvesting and Classifying the Learning Material.   46
      References   48

6   Recommender Systems in E-Learning Environments.   51
      6.1   Recommendations and Recommender Systems   52
      6.2   The Most Important Requirements and Challenges
             for Designing a Recommender System in E-Learning
             Environments   57
      6.3   Recommendation Techniques for RS in E-Learning
             Environments—A Survey of the State-of-the-Art   60
             6.3.1   Collaborative Filtering Approach   61
             6.3.2   Content-Based Techniques   65
             6.3.3   Association Rule Mining.   67
      References   70

**7   Folksonomy and Tag-Based Recommender Systems
      in E-Learning Environments**   77
      7.1   Comprehensive Survey of the State-of-the-Art in Collaborative
             Tagging Systems and Folksonomy   78
             7.1.1   Tagging Rights   80
             7.1.2   Tagging Support   80
             7.1.3   Aggregation   81
             7.1.4   Types of Object.   81
             7.1.5   Sources of Material   82
             7.1.6   Resource Connectivity   82
             7.1.7   Social Connectivity   82
      7.2   A Model for Tagging Activities.   83
      7.3   Tag-Based Recommender Systems   86
             7.3.1   Extension with Tags.   86
             7.3.2   Collecting Tags   87
      7.4   Applying Tag-Based Recommender Systems to E-Learning
             Environments   89
             7.4.1   FolkRank Algorithm.   91
             7.4.2   PLSA.   92
             7.4.3   Collaborative Filtering Based on Collaborative
                      Tagging   95

7.4.4 Tensor Factorization Technique for Tag
        Recommendation   98
7.4.5 Most Popular Tags 106
7.5 Limitations of Current Folksonomy and Possible Solutions 108
References 110

Part III Semantic Web Technologies in E-Learning

8 Semantic Web 115
8.1 Knowledge Organization Systems 116
8.2 Ontologies 117
      8.2.1 Adaptive Educational Systems Technologies
              in E-Learning 119
      8.2.2 Standards for E-Learning Environments 119
      8.2.3 Semantic Web Methodologies 120
      8.2.4 Representation of Ontologies 121
      8.2.5 Development Practices of E-Learning Systems 124
      8.2.6 The Objective of Ontologies 126
      8.2.7 Ontology Application 127
8.3 Semantic Web Languages 129
      8.3.1 XML—eXtensible Markup Language 129
      8.3.2 RDFS—Resource Description Framework Schema 130
      8.3.3 OWL—Ontology Web Language 131
8.4 Graphical Environments for Ontology Development 132
      8.4.1 Protégé 132
      8.4.2 NeOnToolkit 134
      8.4.3 TopBraid Composer 134
      8.4.4 Vitro 135
      8.4.5 OWLGrEd 135
      8.4.6 Knoodl 135
8.5 Educational Ontologies 136
      8.5.1 Domain Ontology 136
      8.5.2 Task Ontology 137
      8.5.3 Teaching Strategy Ontology 137
      8.5.4 Learner Model Ontology 138
      8.5.5 Interface Ontology 138
      8.5.6 Communication Ontology 138
      8.5.7 Educational Service Ontology 139
8.6 Adaptation Rules 139
      8.6.1 Semantic Web Rule Language (SWRL) 140
      8.6.2 Jess 141
8.7 Architecture of Semantic E-Learning Systems 142
References 145

9   **Design and Implementation of General Tutoring**
     System Model   149
     9.1   Architecture of General Tutoring System Model   150
     9.2   System's Ontologies.   151
            9.2.1   Main Components of Ontologies   153
            9.2.2   Domain Ontology   153
            9.2.3   Task Ontology.   156
            9.2.4   Learner Model Ontology.   158
            9.2.5   Teaching Strategy Ontology   161
            9.2.6   Interface Ontology   162
     9.3   Adaptation Rules   163
            9.3.1   Syntax of Adaptation Rules.   163
            9.3.2   Learning Styles Identification.   165
            9.3.3   Rules for Building Learner Model   173
            9.3.4   Adaptation Based on Resource Sequencing   176
     9.4   Course Development   179
     References   181

Part IV   **Case Study: Design and Implementation of Programming**
          Tutoring System

10   **Design, Architecture and Interface of Protus 2.1 System**   185
     10.1   Personalised Programming Tutoring Systems.   186
            10.1.1   Programming Tutoring Systems   186
            10.1.2   Tutoring Systems with Implemented
                     Recommendation   188
     10.2   Previous Versions of Protus 2.1.   189
            10.2.1   Mag System   190
            10.2.2   Protus System   192
     10.3   Protus 2.1   201
            10.3.1   Learner's Interface   201
            10.3.2   User Interface for Teachers and Course
                     Administrators   202
     10.4   Development of Ontologies for Java Programming Course   206
            10.4.1   Domain Ontology   206
            10.4.2   Learner Model Ontology.   208
            10.4.3   Teaching Strategy Ontology   210
            10.4.4   Task Ontology and User Interface Ontology   211
     References   211

11   Personalization in Protus 2.1 System.   213
  11.1   The Protus 2.1 Component for Making Recommendations   214
  11.2   Learning Style Identification in Protus 2.1.   217
      11.2.1   Adaptation Process in Protus 2.1   217
      11.2.2   Calculation of Initial Learning Styles   219
      11.2.3   Adaptation of User Interface Based
            on the Learning Styles   222
  11.3   Resource Sequencing   227
      11.3.1   Identification of Sequences of Learning Activities
            and Personalized Recommendation.   229
  11.4   Recommendation Process Based on Collaborative Filtering   232
  11.5   Tag-Based Personalized Recommendation Using Ranking
        with Tensor Factorization Technique   234
      11.5.1   Generating Initial Tensor.   235
      11.5.2   Computing Tensor Factorization.   235
      11.5.3   Generating a List of Recommended Items   236
      11.5.4   Tag-Based Recommendation in Protus 2.1.   236
  11.6   Use and Functioning of the System   241
      11.6.1   Integration of Java Programming Course
            in Protus 2.1   243
  11.7   Educational Material in Protus 2.1   245
  11.8   Course Organization and Structure   250
      11.8.1   Testing in Protus 2.1   251
      11.8.2   Evaluation Process   256
  References   256

Part V   Evaluation and Discussion

12   Experimental Evaluation of Protus 2.1   261
  12.1   Data Set for Experiment   262
  12.2   Data Clustering   262
  12.3   Statistical Properties of Learners' Tagging History   264
      12.3.1   Learners' Activities   264
      12.3.2   Tag Usage.   265
      12.3.3   Tag Entropy Over Time   266
      12.3.4   Semantic Analysis of Tags   267
  12.4   Experimental Protocol and Evaluation Metrics   268
  12.5   Evaluation of Several Suitable Recommendation
        Techniques   269
      12.5.1   Settings of the Algorithms.   269
      12.5.2   Results of Selected Methods Evaluation   271
  12.6   Expert Validity Survey.   279
  12.7   Evaluation of Protus 2.1 System from the Educational
        Point of View   280
  References   284

13 Conclusions and Future Directions 287
13.1 Contributions of the Monograph 289
13.2 Future Work and Open Research Questions 293
References 294

About the Authors

Aleksandra Klašnja-Milićević, Ph.D. is Assistant Professor at Faculty of Sciences, University of Novi Sad, Serbia. She attended the Faculty of Technical Sciences at the University of Novi Sad, Department of Electrical Engineering and Computer Science, receiving a B.Sc. degree in 2002. She joined the graduate programme in Computer Sciences at Faculty of Sciences, Department of Mathematics and Informatics, University of Novi Sad in 2003, where she received her M.Sc. (2007) and Ph.D. degrees (2013). Her research interests include e-learning and personalization, information retrieval, internet technologies, recommender systems and electronic commerce. She actively participates in several international projects. She has also served as programme committee member of several international conferences. She co-authored one university textbook. She has published 30 scientific papers in proceedings of international conferences and journals.

Boban Vesin, Ph.D. is Research Engineer in the joint Software Engineering Division at the University of Gothenburg and Chalmers University of Technology. Previously, he was Lecturer at Higher School of Professional Business Studies, University of Novi Sad, Serbia and Software Engineer at Schneider Electric DMS Novi Sad, Serbia. He earned his Ph.D. at the Faculty of Science, University of Novi Sad, in 2014. His major research interests are e-learning and personalization in intelligent tutoring systems. He has published over 35 scientific papers in proceedings of international conferences and journals in the field of programming, e-learning, semantic Web and software engineering.

Mirjana Ivanović, Ph.D. since 2002 has been Full Professor at Faculty of Sciences, University of Novi Sad, Serbia. She is member of University Council for informatics. Author or co-author is, of 13 textbooks, five edited proceedings, one monograph and of more than 340 research papers on multi-agent systems, e-learning and Web-based learning, applications of intelligent techniques (CBR, data and Web mining), software engineering education, and most of them are published in international journals and proceedings of high-quality international conferences. She is/was a member of programme committees of more than 200 international conferences and general chair and programme committee chair of several international conferences. Also she has been invited speaker at several international conferences and visiting lecturer in Australia, Thailand and China. As leader and researcher she has participated in numerous international projects. Currently she is Editor-in-Chief of Computer Science and Information Systems Journal.

Prof. Dr. Zoran Budimac since 2004 has been full professor at Faculty of Sciences, University of Novi Sad, Serbia. Currently, he is head of Computing Laboratory and Chair of Computer Science. His fields of research interests involve: software quality assurance, software engineering, distributed programming, programming languages and tools and educational technologies. He has been principal investigator of more than 20 international and national projects. He is author of 13 textbooks and more than 300 research papers most of which are published in international journals and international conferences. Also he has been invited speaker at several international conferences and visiting lecturer at several universities. He is/was a member of programme committees of more than 100 international conferences and is member of editorial and managing boards of "Computer Science and Information Systems Journal."

Lakhmi C. Jain, Ph.D. is with the Faculty of Education, Science, Technology, and Mathematics at the University of Canberra, Australia and Bournemouth University, United Kingdom. He is a Fellow of the Institution of Engineers Australia. Professor Jain founded the KES International for providing a knowledge exchange, cooperation and teaming.

www.kesinternational.org

Involving around 5,000 researchers drawn from universities and companies worldwide, KES facilitates international cooperation and generate synergy in teaching and research. KES regularly provides networking opportunities for professional community through one of the largest conferences of its kind in the area of KES. His interests focus on the artificial intelligence paradigms and their applications in complex systems, security, e-education, e-healthcare, unmanned air vehicles and intelligent agents.

Abbreviations

ABSS	Agent-Based Search System
CBR	Case-Based Reasoning
CF	Collaborative Filtering
CSA	Cognitive Styles Analysis
CSL	Cognitive Styles of Learning
EM	Expectation–Maximization
FSLSM	Felder–Silverman Learning Style Model
GEFT	Group Embedded Figures Test
HBDI	Herrmann Brain's Dominance Instrument
HOSVD	Higher Order Singular Value Decomposition
ICT	Information and Communication Technology
IDE	Integrated Development Environment
IF	Information Filtering
ILS	Index of Learning Styles
IMS CP	IMS Content Packaging
IMS SS	IMS Simple Sequencing Specification
JDBC	Java DataBase Connectivity
JECA	Java Error Correction Algorithm
LA	Learning Activities
LAO	Learning Application Objects
LCMS	Learning Content Management Systems
LMS	Learning Management Systems
LO	Learning Objects
LOM	Learning Object Metadata
LOs	Learning Objects
LSA	Latent Semantic Analysis
LSI	Learning Style Inventory
LSP	Learning Styles Profiler
LSQ	Learning Style Questionnaire
MSD	Gregorc Mind Styles Delineator

MSP	Motivational Style Profile
OWL	Web Ontology Language
PLE	Personal Learning Environment
PLSA	Probabilistic Latent Semantic Analysis
QTI	Question & Test Interoperability
RACOFI	Rule-Applying Collaborative Filtering
RDF	Resource Description Framework
RS	Recommender Systems
RTF	Ranking with Tensor Factorization
SCORM	Sharable Content Object Reference Model
SPARQL	Simple Protocol and RDF Query Language
SVD	Singular Value Decomposition
SWRL	Semantic Web Rule Language
W3C	World Wide Web Consortium
WCR	Web Content Resources
XML	Extensible Markup Language

Abstract

Semantic Web is a next generation of Web that is trying to present information in such a way that they can be used by computers, for display, automation, integration and reuse among different applications. The aim of the monograph is to present the design, implementation and real-life evaluation of a tutoring system for maintenance of courses from various domains using semantic Web technologies. This process includes the creation of the fundamental building blocks of ontologies and rules for carrying out the actions for adaptation of teaching materials and learning processes

The subject of the monograph includes the implementation of a conceptual model of tutoring system for e-learning in different domains using semantic Web technologies and implementation of a prototype system that is applied in designing a personalized tutoring system for learning the Java programming language basics.

Part I
Preliminaries

Chapter 1
Introduction to E-Learning Systems

Abstract Recently e-learning systems are experiencing rapid development. The advantages of learning through a global network are manifold and obvious: the independence of time and space, learners can learn at their own pace, learning materials can be organized in one place and used-processed all around the world. One of the most important segments in today's development and use of the e-learning system is the personalization of content and building of user profiles based on the learning behaviour of each individual user. The personalization options increase efficiency of e-learning, thus justifying the higher initial cost of their construction. In order to personalize the learning process and adapt content to each learner, e-learning systems can use strategies that have the ability to meet the needs of learners. Also, these systems have to use different technologies to change the environment and perform the adaptation of teaching materials based on the needs of learners. The process of adaptation can be in the form of adaptation of content, learning process, feedback or navigation. This chapter introduces the motivation and objectives studied in the subsequently presented research, and presents major standards and specifications in e-learning.

Recently e-learning systems are experiencing rapid development. The advantages of learning through a global network are manifold and obvious: the independence of time and space, learners can learn at their own pace, learning materials can be organized in one place and used-processed all around the world. E-learning is therefore proved to be efficient, flexible and affordable. Development of e-learning capacities is obviously much more demanding and more expensive than the development of a static system since it is desirable for the system to provide different forms of the same teaching materials which are needed for the successful implementation of personalized learning. However, the personalization options increase efficiency of e-learning, thus justifying the higher initial cost of their construction.

One of the most important segments in today's development and use of the e-learning system is the personalization of content and building of user profiles based on the learning behaviour of each individual user. The constructed profile is

© Springer International Publishing Switzerland 2017

A. Klašnja-Milićević et al., *E-Learning Systems*,

Intelligent Systems Reference Library 112, DOI 10.1007/978-3-319-41163-7_1

intended to help the system in the selection of content and information presented to the user at a given moment. In order to personalize the learning process and adapt content to each learner, e-learning systems can use strategies that have the ability to meet the needs of learners. Also, these systems have to use different technologies to change the environment and perform the adaptation of teaching materials based on the needs of learners. The process of adaptation can be in the form of adaptation of content, learning process, feedback or navigation.

1.1   Web-Based Learning

Web-based learning involves all aspects of the learning process that use the World Wide Web as a basic technology and the medium of communication. Other terms are also used such as 'Online learning', 'Virtual education', 'Internet-based learning' or 'Education via computer-mediated communication' (Devedzic 2004).

Basic characteristics of Web-based learning (Harsh and Sohail 2002) are:

- separation of teachers and learners (that is the main difference of this type of learning compared to the traditional),
- the use of Web technologies for presentation and distribution of educational content,
- possibilities for two-way communication among learners and between learners and teachers.

Since 1990, Web-based learning has become an important branch of education. For learners it offers virtually unlimited access to information and knowledge. It also offers customization of courses to each individual learner, 'telelearning', the possibilities of mutual cooperation among learners and clear benefits of classroom and platform independence (Brusilovsky 2004).

On the other hand, teachers and course authors may use numerous benefits of online courses, such as 'teleteaching', authoring tools for course development, inexpensive and efficient data storage and distribution of course materials, digital libraries, etc.

There are a number of important elements associated with Web-based learning, such as: e-learning, distance learning and personalized learning. The following sections include further clarified certain concepts and related technologies.

1.2   E-Learning

The term "e-learning" has been in existence since October 1999 when it was used during a CBT Systems seminar in Los Angeles. Together with the terms "online learning" and "virtual learning", this word was meant to qualify "a way to learn based on the use of new technologies allowing access to online, interactive and sometimes personalized training through the Internet or other electronic media (intranet, extranet, interactive TV, CD-ROM, and so on), so as to develop competencies while the process of learning is independent from time and place" (Barth et al. 2014).

By some definitions, e-learning includes not only the Internet as a technical support to learning, but also other media and resources such as Intranet, audio and video discs, satellite broadcasting of lectures, interactive television, wireless and mobile devices, and so on (Devedzic 2006). However, e-learning is converted into Internet learning as it primarily uses Internet technologies for the creation, adoption, transfer and facilitation of the learning process (Dutta 2006). One of the objectives of e-learning is the development of individualized, understandable and dynamic learning content in real time.

E-learning can be realized in or out of the classroom. It can be self-paced, asynchronous learning or it may also be instructor-led, synchronous learning. E-learning usually stands for distance flexible learning. Except that e-learning can be used in blended learning form, i.e. in conjunction with traditional face-to-face teaching. We are nowadays witnessing the emergent contemporary new technologies and methods that make big differences and challenges in all kinds of educational settings.

On the other hand, the principles behind e-learning and early forms of e-learning existed even in the 19th century. Long before the appearance of the internet, distance courses on particular subjects were being offered to students from different countries. Isaac Pitman, a qualified teacher, taught the pupils shorthand via correspondence in the 1840s. In fact Pitman was sent completed assignments by his students via the mail system and (s)he would then send them more work to be finished.

The first testing machine, allowing students to test their knowledge, was invented in 1924. In 1954, B.F. Skinner, a Harvard Professor, invented the "teaching machine". This machine enabled schools to administer programmed instruction to the students. The first computer based training program, starting from University of Illinois, was PLATO-Programmed Logic for Automated Teaching (Murphy and Appeal 1978). PLATO was equipped with many modern concepts in multi-user computing like: e-mail, instant messaging, chat rooms, forums, message boards, online testing, picture languages, remote screen sharing, and multiplayer games.

The first e-learning systems served only for delivering information to students. Starting from the 70s they became more interactive. In Britain the Open University (Gourley and Lane 2009) highly organized their activities in order to take advantages of e-learning and they primarily have been focused on learning at a distance. With the appearance of the internet the Open University began to offer a variety of interactive educational facilities and faster correspondence with students via email and so on.

The revolution in e-learning happened when first MAC computers in the 1980s enabled individuals to have computers in their homes. This possibility offered students an easier way to learn about particular subjects and develop certain skill sets.

E-learning nowadays is one important issue to spread all forms of education in our society. E-learning tools and delivery methods expanded as a consequence of the introduction of the computers and internet in the late 20th century. Also virtual learning environments began to truly progress, with people gaining access to a richness of online information and e-learning opportunities.

At the end of the 90s the first learning management systems (LMS) started to be used frequently. Some universities designed and developed their own systems, but most of the educational institutions started with systems off the market. One of the key commercial systems was offered by the American company Blackboard (Bradford et al. 2007). In the meanwhile a lot of LMSs have been developed and appeared. Today in academia predominantly open-source MOODLE system (Aranda 2012; Horvat et al. 2015) has been in use all over the world. Such systems offer to students and teachers wide functionalities: exchange learning materials, do tests, communicate with each other in many ways, track and trace the progress, and so on.

Another important step and achievement in e-learning area happened in the 2000s. In that period a lot of companies all over the world began using e-learning to train their employees. Workers have had the opportunity to improve upon their industry knowledge base and expand their skills. At home individuals can access programs that offered them the ability to earn online degrees and expand their knowledge. Today, e-learning is more popular than ever and appears in different forms and offers unlimited opportunities and possibilities to universities, companies, and individuals in the form of formal and informal learning.

Basically e-learning implies that the learner is at a distance from the tutor/instructor and uses some form of technology (usually a computer) to access the learning material. Further the learner uses technology to interact with the tutor/instructor but also with other learners. Crucially e-learning refers to the use of information and communication technology (ICT) to enhance and/or support learning. This certainly encompasses a range of systems, starting from students using e-mail and accessing course materials online to the whole programs delivered online. So e-learning is a completely new education philosophy via the Internet, network, or standalone computer. E-learning applications and processes include web-based learning, computer-based learning, virtual classrooms and digital/virtual collaboration.

In contemporary education e-learning supports the different phases of traditional learning and sometimes it is the only possible method of teaching (e.g. impaired students, absence of teaching structures, and so on). From this point of view, it is important to define educational structure and contextualize and tailor it having in mind:

- teachers (tutors, instructors) with their personal teaching approaches, and
- students (learners, workers) with their personal studying manner.

On contrary to the traditional teaching methods followed a "one size fits all" approach. Recently it has become clearer that different people learn in different ways. It was widely recognized that a personalized approach can improve the learning process, helping people becoming effective lifelong. Personalization helps learners in developing a feeling of competence and autonomy as they are trusted with the management of their own learning process.

There are two main directions for the development of e-learning: technical and pedagogical. Many authors emphasize the technology. Others use technology only as a tool for presenting content, with an emphasis on the various approaches for the presentation of the course material to the learners. For them, e-learning is essentially just learning. In the course development, the emphasis is on explanation of how people learn, how acquire skills and take on the information, what are their learning styles, and so on. Only after these primary aspects, the question arises, how the electronic presentation of the material can be tailored to the learner.

E-learning is usually presented as an interaction between the learners and the simulated electronic environment associated with the domain knowledge interesting for learners. This electronic environment, in the case of online learning, may be the Internet (in different versions), intranet or various electronic media. In all these cases, efficient environment for learning is offered through the interactive use of text, images, audio and video material, animations and simulations. It may also include the entire virtual environment. It is possible to learn independently or in a group and also independently determine the appropriate pace of progress.

E-learning has many advantages over traditional learning in the classroom. Primarily, the pace of e-learning is tailored to the learner. Then, the costs are usually lower, without time and space limitations. Also, the material is easier to maintain. A good alternative to regular e-learning is *Blended learning* which is a combination of traditional and e-learning. This form of learning involves combining traditional classroom lectures with occasional/parallel teaching using some form of tutoring systems or systems for e-learning (Garrison and Vaughan 2008).

1.3 E-Learning Objects, Standards and Specifications

E-learning as majority of other ICT areas is oriented to "civilizing stage" in order to develop common technical interoperability standards and specifications. Standards are essentially abstract and they impose only structural limits on what can be done with them. In fact, they are intended to provide a sustainable means for the practical use of appropriate technologies and ways of working.

E-learning standards and specifications have to establish common patterns to support different aspects of educational and meta-educational activities and processes.

There are a lot of widely recognized standards in eLearning which describe different sub-areas. Some of them are devoted to content description, or to the sequencing of items to be delivered or describe learner and content packaging and so on.

Also, there exist many other standards underpinning, which support the use of technology in educational processes like: media standards (USB or CD/DVD-ROM), network standards (TCP/IP), and file standards (HTML, XML, and so on). Also a new category called meta-standard has been introduced, like: quality assurance of educational processes, legal standards as laws and licenses, and political standards as policies and procedures.

1.3.1   E-Learning Objects

For the success of e-learning, the most important factor is a suitable organization of the material that is taught in a way that corresponds to the interactive electronic presentation materials (Vesin and Ivanović 2004). The worst solution is to copy the traditionally written material in the files and simply present them on the screen to learners. In contrast, preparation of e-learning courses is a long process that requires a lot of effort by course creators and the entire team working on the project. Educational institutions and universities that offer online courses have whole departments that are in charge of organizing representation and presentation of e-learning material.

There are a few simple rules of organizing e-learning materials. They are all consequences of the general pedagogic rules and requirements for human-computer interaction:

- It is important to accurately define the target users (learners and their level of knowledge) and the objectives and outcomes of the course (i.e. what knowledge is expected that learners acquire after the completion of the course). In cases where the course is conducted through networks, organization of material on the server side is only one of the problems—it is necessary to have in mind the hardware that the learner use, as well as possible data transfer speed in order to provide satisfactory communication in real time.
- It is necessary to divide the teaching material into modules such as chapters or lessons, which will enable learners to gain insight into the general structure of the material, acquire the aims of the course, and to more easily follow the details in chapters and lessons.

There are wide ranges of user tools that help instructors to prepare material for the above mentioned instructions (Devedzic 2004).

Learning and teaching activities are deeply connected to the forms and ways of representation and presentation of teaching material. The absence of agreed unique standard in this domain as a consequence brings a lot of different interpretations and definitions of learning objects as the basic unit for representations of teaching material.

For learning objects (LO) variety of names has been used, including content objects, chunks, educational objects, information objects, intelligent objects, knowledge bits, knowledge objects, learning components, media objects, reusable curriculum components, nuggets, reusable information objects, reusable learning objects, testable reusable units of cognition, training components, units of learning and so on.

Nevertheless, learning objects have been frequently defined as small, unique, compact and reusable pieces (units, entities) of teaching material (text, video or audio representation, animation, interactive simulation, interactive exercises) (Pitkänen and Silander 2004). IEEE Learning Technology Standards Committee, recently issued a new definition of LO as "any entity, digital or non-digital, that may be used for learning, education or training" (Committee et al. 2002; González and Ruggiero 2009). According to this and other similar definitions, LOs are, generally speaking, digital elements that are used in learning processes and consist of following parts: texts, pictures/images, digital video and audio records, interactive multimedia, tests, and lessons. Combining different LOs teachers produce teaching material for whole courses. The common understanding is that LOs possess significant potential as basic building blocks of the wide spectrum that can be used in different technology enhanced learning systems and environments.

LOs could be classified based on different criteria, but always is necessary to have in mind their pedagogical aspects and wider context of usage. There is a set of obvious criteria that have to be taken into account for assessment of flexibility and quality of the LO:

- pedagogical neutrality,
- reusability potential,
- possibility for personalization
- media independence.

Learning object design raises issues of portability, and of the object's relation to a broader learning management system. In Committee et al. (2002) reusability of existing teaching materials in new systems and contexts has been considered. Relationship between the degree of reusability and complexity of teaching material and LOs is presented in Fig. 1.1.

It can be seen that as LOs are getting more and more complex their relation and connection to the particular context (they are created for) has been increasing. Accordingly, the reusability potential of LOs is decreasing as LOs getting more and more complex.

Based on different sources analysed in Zdravkovaet al. (2012) essential characteristics of ideal reusable LO are pointed out:

- modular, self-confident, interoperable between different e-learning environments;
- non-sequential;
- able to satisfy particular teaching goal;
- widely and easily available, easily modifiable;
- rather simple and characterized by a small set of tags;
- independent on particular format of an e-learning system.

LO and its metadata can include some additional types of information:

- General Course Descriptive Data: subject area, language of content, course identifiers, descriptive keywords, descriptive text,

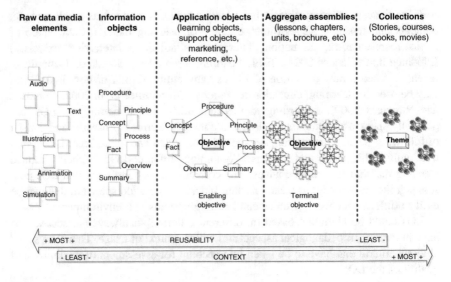

Fig. 1.1 Reusability and complexity of teaching material

- Instructional Content: text, images, sound, video, web pages,
- Typology: presentation, simulation, practice, information, and contextual representation, conceptual models,
- Relationships to Other Courses,
- Life Cycle: version, status,
- Assessments and Quizzes: questions, answers,
- Glossary: definition, terms, acronyms,
- Rights: restrictions on use, copyrights, cost,
- Educational Level: grade level, typical learning time, age range, and difficulty.

It is essential that LOs could be identified by content management systems and search engines. To achieve better visibility and accessibility of LOs appropriate descriptive learning object metadata has to be used. The typical, most important pieces of metadata include:

1. **Topic**—usually represented in a taxonomy, determines the subject which is instructed in the learning object,
2. **Objective**—the educational objective,
3. **Interactivity**—the interaction model of the learning object,
4. **Prerequisites**—the list of skills which the learner must possess before using the learning object,
5. **Technology requirements**—the system requirements, necessary that learning object has to be presented to the learner.

Creation of high-quality, usable and efficient LOs is definitely not an easy task. So, to obtain a high level of interoperability and reusability of teaching materials essential activities have to be oriented to standardization.

1.3.2 E-Learning Specifications and Standards

One of the main motivations and efforts to establish e-learning standards and specifications were to assure synchronous and asynchronous exchange of educational materials between different systems. Asynchronous exchange over time assumes that any one technology can be disposed of and/or replaced while keeping the investment in the content and processes they support.

To improve reusability of the content, the educational material is naturally organized in small self-contained pieces of information, i.e. learning objects. Specific kind of such systems that manage LO are called Learning Content Management Systems (LCMS). LO is transportable and reusable pieces of instruction that are digitally delivered and managed. Several organizations (IEEE, AICC, IMS, and ADL) have developed standards and specifications as guidelines and best practices on the description and use of e-learning content. There are a lot of the proposed specifications and the most prominent could be divided into three groups: metadata, content package and educational design (Verdú et al. 2012).

Metadata—A learning object consists of one or more educational resources that are described by metadata. One of the most used metadata standards is the LOM—Learning Object Metadata (LOM), a double IEEE and IMS standard. LOM is usually encoded in XML, used to describe a learning object. LOM appeared as a reaction to the Dublin Core standard that was characterized as too simple for adequately describing learning resources. The purpose of LOM is multifold: to facilitate learning object interoperability, to support their reusability, and to aid discoverability. LOM is also integrated in IMS Content Packaging (IMS CP) standard.

Another well known metadata specification is the Dublin Core Metadata, which provides a simpler set of elements useful for sharing metadata across heterogeneous systems. Dublin Education Working Group is still working on improvements of the Dublin Core for the specific educational needs.

Content Package—Content Packaging is an essential activity for storing of e-learning material and reusing it in different systems. IMS Content Packaging (IMS CP) is one of the most widely used formats.

There are other package specifications, mostly derived from other systems as application profiles. Sharable Content Object Reference Model (SCORM) is a widely well known content packaging specification. It extends IMS CP with more sophisticated sequencing and Contents-to-LMS communication.

Educational Design—Learning objects are usually organized in items. Such organization determines a path through the items.

The IMS CP specification proposes a manifest section called Organizations. Organizations are used to design pedagogical activities and specify the sequencing of instructions. It is usually arranged as tree-based structure of learning items pointing to the resources included in the package. Some other standards exist, such as IMS Simple Sequencing (IMS SS) and IMS Learning Design (IMS LD). They provide to the teachers mechanisms for coordination of the educational instructions founded on students' profile making the instruction more flexible.

In spite the fact that educational community pays a lot of attention to the development of this kind of standards the design of more complex adaptive behaviour is still hard to achieve.

In the rest of the subsection we will briefly present some of the most important educational standards.

1.3.2.1 S1. IEEE LOM and IMS Learning Resource Metadata

The IEEE LOM as a standard conceptual model has been developed by working group IEEE 12 (Learning Object Metadata Working Group). It is used to specify the syntax and semantics and to describe an LO and similar digital resources used to support learning. Apart from that it aims to provide an extensive metadata description for LOs. In the context of online LMSs the rationale of the IEEE LOM is to support the reusability and facilitate their interoperability. This standard consists of several categories (Verdú et al. 2012):

- **General**—context independent and semantic properties of the LO,
- **Lifecycle**—properties related to the description of the resource's lifecycle,
- **Meta-metadata**—information about the metadata itself,
- **Technical**—technical properties,
- **Educational**—learning and pedagogical properties of the LO,
- **Rights**—conditions for the use of the LO and intellectual property rights,
- **Relation**—the resource's relationship to other LOs,
- **Annotation**—comments on the educational use of the LO,
- **Classification**—description of this LO relative to a particular classification system.

These main categories are further divided into nearly sixty fields. From the IEEE LOM Schema working document the new standard has been derived—the IMS Learning Resource Metadata. The IMS also provides a best practice implementation guide and XML bindings.

Fortunately, LOM was designed in an extensible manner. Some typical ways to extend the LOM model are following:

- combining the LOM elements with elements from other specifications;
- defining extensions to LOM elements while preserving its set of categories;
- simplifying LOM, reducing the number of LOM elements and its choices;

- extending and reducing at the same time the number of LOM elements.

Based on the previously mentioned extensions, the IMS GLC created the Question and Test Interoperability (QTI) specification. This specification offers a data model for the representation of different kinds of questions (e.g. multiple response, multiple choice, short text questions and fill in-the-blanks) and tests data and their corresponding results reports.

1.3.2.2   S2. Dublin Core Metadata Initiative

At a workshop held in 1995 in Dublin, Ohio, United States the Dublin Core (DC) standard has been proposed. The DC metadata element set is a standard for describing primarily web information resources. It consists of simple sets of elements to facilitate describing, sharing, finding, and managing information. The fifteen basic DC metadata elements exist: Contributor, Coverage, Creator, Date, Description, Format, Identifier, Language, Publisher, Relation, Rights, Source, Subject, Title and Type. The DC has been quickly adopted by many international and interdisciplinary communities. As a consequence of its simplicity qualifiers have been introduced to further specify existing DC elements, in order to increase the precision of the encoded metadata. Two groups of qualifiers exist: element refinement and encoding scheme. An important characteristic of this standard is its aiding to resource discovery and facilitating interoperability. This feature qualifies it as a preferred format in the Open Archives Initiative Protocol for Metadata Harvesting.

1.3.2.3   S3. IMS Learner Information Package

LIP—IMS Learner Information Packaging is also specified by IMS Global Learning Consortium. It includes a collection of information about a product of learning content (creators, providers or vendors) and a learner as well. LIP addresses the interoperability of learner information between different learning management systems and/or Internet-based systems. The core structures of the IMS LIP are based on: accessibilities, activities, affiliations, competencies, goals, identifications, interests, qualifications, certifications, licenses, relationship, security keys, and transcripts. A similar specification, i.e. IEEE Public and Private Information Specification have been proposed by IEEE.

1.3.2.4   S4. IMS Content Packaging

IMS Content Packaging Specification, as an interoperability specification, allows tools for content creation, LMSs, and runtime environments to share content in a standardized set of structures. The main purpose of this specification is: to standardize the way of definition of learning resources and to facilitate the organization of different components related to learning content. This obtains structuring learning content into a package and in this way supports interoperability, i.e. exchange of packages between different learning environments.

An IMS CP learning object assembles resources and meta-data into a distribution medium, usually an archive in ZIP format, with its content described in a manifest file in the root level. The manifest file adheres to the IMS CP schema and contains specific sections (Verdú et al. 2012):

- Metadata—describes the package as a whole;
- Organizations—describes the organization of the content within a manifest;
- Resources—contains references to resources (files) needed for the manifest and metadata describing these resources;
- Sub-manifests—defines sub packages.

To describe the learning resources included in the package the manifest uses the LOM standard. In meanwhile, IMS Global Learning Consortium issued the IMS Common Cartridge that supports some other standards (e.g. IEEE LOM, IMS CP, IMS QTI, IMS Authorization Web Service). Their main intention is playing an important role in the organization and distribution of digital learning content. The IMS CC manifest includes references for two types of resources:

- Web Content Resources (WCR): static web resources such as PDF documents, GIF/JPEG images, HTML files, and so on.
- Learning Application Objects (LAO): special kind of resource types that need to be additionally processed before they can be used within the target system. QTI assessments, Discussion Forums, Web links, Basic LTI descriptors, etc. are typical examples of LAO.

1.3.2.5   S5. IMS Simple Sequencing

IMS Learning Design (IMS LD) specification is a meta-language for describing pedagogical models and educational goals through enabling the modelling of learning processes (Leo et al. 2004). There are several IMS LD-aware tools in form of players (e.g. CopperCore, .LRN) and authoring/export tools (e.g. Reload, LAMS) (Dalziel 2006).

The IMS Simple Sequencing Specification (IMS SS) is used to describe paths through a collection of learning activities. The specification determines the order in which learning activities have to be presented to a learner. Apart from that it declares the conditions under which a resource is delivered during an e-learning instruction.

In fact, this specification describes behaviours and functionality that conforming systems must implement. It includes definitions of rules that describe learner's interactions with content and the sequencing of LO based on instructional design strategies selected by the course's instructor. The LMS maintains a simple user model based on both the LOs the user has visited and intermediate test results. The LMS then decides what LO should be next in the sequence. IMS SS is labelled as simple because it supports a limited number of widely used sequencing behaviours so some are ready to label it as a simple mechanism. The wide range of possible sequencing mechanisms are not addressed in this specification, like: schedule-based sequencing, artificial intelligence based sequencing, collaborative learning, customized learning, sequencing requiring data from closed external systems (e.g., sequencing of embedded simulations), or synchronization between multiple parallel learning activities. IMS SS is a part of SCORM standard.

1.3.2.6 S6. ADL SCORM

The SCORM (Sharable Content Object Reference Model) has been developed by the Advanced Distributed Learning Initiative, under the umbrella of the United States Department of Defense, as a set of standards and specifications for e-learning (de Oliveira and Gomes 2015). It is the de facto industry, purely a technical standard for e-learning interoperability. Interoperability, accessibility, and reusability of learning content as primary goal have been enabled by packaging content in a zip form. Additional capabilities like sequencing, which includes rules that specify the order in which a learner may experience LOs, been included in SCORM 2004. The standard is an XML-based, and it utilizes a significant work performed by several sources IEEE,[1] AICC (CBT),[2] IMS Global Learning Consortium,[3] and Ariadne Foundation.[4] In fact SCORM tells programmers how to write their code so that it can "play well" with other e-learning software.

The next generation of SCORM is happening nowadays. It's called the Tin Can API[5] and represents a huge leap forward for the e-learning community.

1.3.3 Analysis of Standards and Specifications

Standards for content and descriptions of learner's information are becoming increasingly important in LMSs. Most of them are developed for use by LMSs. Standards make possible exchanging content and learner information and interoperability between different learning environments. They can be used independently of the underlying technology of the specific environment as they are usually

---

[1]http://www.ieee.org/index.html.

[2]http://adlnet.gov/.

[3]http://www.imsglobal.org/.

[4]http://www.ariadne-eu.org/.

[5]http://scorm.com/tincanoverview/.

described in a platform-neutral manner (predominantly in XML). Therefore, it could happen that their use in an adaptive e-learning environment is restricted in a way that a specification might not be sufficient to support adaptivity. However, the baseline support for standards is imperative and represents a starting point for interoperability between different systems. Today, dozens of LMSs will be available in academia, open source community and commercial market. For that reason, specifications like SCORM enable communication with different LMSs and the dissemination of learning material.

References

Aranda, A. D. (2012). Moodle for distance education. *Distance Learning, 8*(2), 25–28.

Barth, M., Adomßent, M., Fischer, D., Richter, S., & Rieckmann, M. (2014). Learning to change universities from within: A service-learning perspective on promoting sustainable consumption in higher education. *Journal of Cleaner Production, 62*, 72–81.

Bradford, P., Margaret Porciello, N., & Balkon, D. B. (2007). The blackboard learning system. *The Journal of Educational Technology Systems, 35*, 301–314. http://doi.org/10.2190/X137-X73L-5261-5656

Brusilovsky, P. (2004). KnowledgeTree: A distributed architecture for adaptive e-learning. In *WWW Alt. '04: Proceedings of the 13th International World Wide Web Conference on Alternate Track Papers & Posters* (pp. 104–113). http://doi.org/10.1145/1013367.1013386

Committee, L. T. S., et al. (2002). IEEE standard for learning object metadata. *IEEE Standard, 1484*(1), 2004–2007.

Dalziel, J. R. (2006). Lessons from LAMS for IMS learning design. In *Sixth International Conference on Advanced Learning Technologies*, (Vol. 3, 1101–1102). http://doi.org/10.1109/ICALT.2006.1652643

de Oliveira, F., & Gomes, A. S. (2015). A development model of units of learning for multiple platforms. In *2015 10th Iberian Conference on Information Systems and Technologies (CISTI)* (pp. 1–6).

Devedzic, V. (2004). Education and the semantic web. *International Journal of Artificial Intelligence in Education, 14*, 39–65.

Devedzic, V. (2006). *Semantic web and education* (Vol. 11). http://doi.org/10.1007/978-0-387-35417-0

Dutta, B. (2006). *Semantic web based e-learning*. Bangalore: Documentation Research and Training Centre Indian Statistical Institute.

Garrison, D. R., & Vaughan, N. D. (2008). *Blended learning in higher education: Framework, principles, and guidelines. booksgooglecom* (Vol. 1st).

González, L. A. G., & Ruggiero, W. V. (2009). Collaborative e-learning and learning objects. *IEEE Latin America Transactions, 7*(5), 569–577. http://doi.org/10.1109/TLA.2009.5361195

Gourley, B., & Lane, A. (2009). Re-invigorating openness at The Open University: The role of open educational resources. *Open Learning: The Journal of Open and Distance Learning*, 37–41. http://doi.org/10.1080/02680510802627845

Harsh, O. K., & Sohail, M. S. (2002). Role of delivery, course design and teacher-student interaction: Observations of adult distance education and traditional on-campus education. *The International Review of Research in Open and Distributed Learning, 3*(2).

Horvat, A., Dobrota, M., Krsmanovic, M., & Cudanov, M. (2015). Student perception of Moodle learning management system: A satisfaction and significance analysis. *Interactive Learning Environments, 23*(4), 515–527.

Leo, D. H., Perez, J. I. A., & Dimitriadis, Y. A. (2004). IMS learning design support for the formalization of collaborative learning patterns. In *IEEE International Conference on Advanced Learning Technologies, 2004. Proceedings.* (pp. 350–355). http://doi.org/10.1109/ICALT.2004.1357434

Murphy, R. T., & Appeal, L. R. (1978). Evaluation of the PLATO IV computer-based education system in the community college. *ACM SIGCUE Outlook, 12*(1), 12–28.

Pitkänen, S. H., & Silander, P. (2004). Criteria for pedagogical reusability of learning objects enabling adaptation and individualised learning processes. In *IEEE International Conference on Advanced Learning Technologies, 2004. Proceedings* (pp. 246–250).

Verdú, E., Regueras, L. M., Verdú, M. J., Leal, J. P., de Castro, J. P., & Queirós, R. (2012). A distributed system for learning programming online. *Computers & Education, 58*(1), 1–10. http://doi.org/10.1016/j.compedu.2011.08.015

Vesin, B., & Ivanović, M. (2004). Modern educational tools. In *Proceedings of PRIM2004, 16th Conference on Applied Mathematics, Budva, Montenegro* (pp. 293–302).

Zdravkova, K., Ivanović, M., & Putnik, Z. (2012). Experience of integrating web 2.0 technologies. *Educational Technology Research and Development.* http://doi.org/10.1007/s11423-011-9228-z

Part II
E-Learning Systems Personalization

Chapter 2
Personalization and Adaptation in E-Learning Systems

Abstract Personalization is a feature that occurs separately within each system that supports some kind of users' interactions with the system. Generally speaking term "Personalization" means the process of deciding what the highest value of an individual is if (s)he has a set of possible choices. These choices can range from a customized home page "look and feel" to product recommendations or from banner advertisements to news content. In this monograph we are interested in personalization in educational settings. The topic of personalization is strictly related to the shift from a teacher-centred perspective of teaching to a learner-centred, competency-oriented one. Two main approaches to the personalization can be distinguished: user-profile based personalization and rules-based personalization. In the first case this is the process of making decisions based upon stored user profile information or predefined group membership. In the second case this is the process of making decisions based on pre-defined business rules as they apply to a segmentation of users. This chapter presents the most popular adaptation forms of educational materials to learners.

Today, personalization is feature that occurs separately within each system that supports some kind of users' interactions between user and the system. Generally speaking term "Personalization" means the process of deciding what the highest value of an individual is if (s)he has a set of possible choices. These choices can range from a customized home page "look and feel" to product recommendations or from banner advertisements to news content.

The concept of "Personalization" can easily be understood taking a closer look at some of the widely existing and using digital technologies that offer personalization and customization options: browser that help to roam the Internet, email and messaging systems, the digital boxes that help to watch TV online.

Two main approaches to the personalization can be distinguished: user-profile based personalization and rules-based personalization. In first case this is the process of making decisions based upon stored user profile information or predefined group membership. In the second case this is the process of making decisions based on pre-defined business rules as they apply to a segmentation of users.

© Springer International Publishing Switzerland 2017
A. Klašnja-Milićević et al., *E-Learning Systems*,
Intelligent Systems Reference Library 112, DOI 10.1007/978-3-319-41163-7_2

2.1 Personalization and Personalized Learning

In this monograph we are interested in personalization in educational settings. In e-learning area, "personalization" has a wide range of new meanings. One of the best explanations could be that "Personalized learning is the tailoring of pedagogy, curriculum and learning environments to meet the needs and learning styles of individual learners" (Baguley et al. 2014).

The topic of personalization is strictly related to the shift from a teacher-centered perspective of teaching to a learner-centered, competency-oriented one. In contrary to conventional e-learning which tends to treat learners as a homogeneous entity, personalized e-learning recognizes learners as a heterogeneous mix of individuals.

Essentially personalized e-learning offers to learner's customization of a variety of the elements of the online education process:

- The learning environment—content and its appearance to the learner (like backgrounds, themes, font sizes, colours, and so on)
- The learning content itself—multimedia representations (like textual, graphical, audio, video, and so on)
- The interaction—include facilitator, student and the learning content (e.g. mouse, keyboard, tap/swipe; e.g. using Quizzes, Online discussions, "Gaming", Tutorials, Adaptive learning approaches)

Apart from the above mentioned ways of personalization (like the "preferences" and "settings" options that most digital tools offer) other aspects of the learning environment and process can be personalized:

- What content should be delivered during the learning process?
- How the content should be delivered with special attention to the sequence of its delivery.
- How students will be evaluated and also with special attention what feedback options will be used.

Nowadays it is unavoidable demand that educators have to re-evaluate e-learning courses and there are a lot of important factors that determine it, like: age, cultural background, the level of education, demographics and so on. Numerous important aspects should be taken into account when deciding to personalize an e-learning environment:

- **Personalize the environment**—determine how online e-learning environments should look like.
- **Personalize the content**—incorporate content from the learners' personal environment (reflect learners' browsing habits and preferences).
- **Personalize the media**—according to their learning styles and preferences some learners like to watch a short video or read a printed PDF files.
- **Personalizing learning sequences**—nonlinear presentation of contents allows learners to choose how they will learn.

- **Personalize the roles using photographs and pictures**—use a photograph of the instructor to make the content more "personal."
- **Personalize the conversation**—use text or voice/video and adjust used sentences.
- **Personalize the navigation**—allow learners to explore various parts of the content.
- **Personalize the learner**—Make the course personal to the learner.
- **Recognize individual competency**—skip known parts of teaching material and start learning the new topics.
- **Personalizing learning objectives**—Enable learners to achieve better the learning objectives.

Harmonization of mentioned aspects will obtain a truly Personal Learning Environment (PLE) and give learners the chance to learn what they want when they want, and even to learn according to the preferred method of learning!

2.2 Adaptation of E-Learning Systems

In the last few decades "Adaptation in E-learning" has generated tremendous interest among researchers in computer-based education. As a consequence, two key terms appeared: adaptivity and adaptability. Adaptivity is such kind of behaviour where the user triggers some actions in the system that guides the learning process, i.e. modifies e-learning lessons using different parameters and a set of pre-defined rules. Adaptability is such kind of behaviour where the user makes changes and takes decisions on the learning process, i.e. it is a possibility for learners to personalize an e-learning lesson by themselves (Khemaja and Taamaallah 2016).

These terms caused a series of possibilities, from those centered on the machine (adaptivity) to those centered on the user (adaptability). Adaptation in e-learning today incorporates new technologies and ways of expression practically moving ahead from Computer Based Training and Adaptive Hypermedia Systems.

Adaptation is usually focused on the student. Also, it is possible adaptation that involves instructors, but it requires deeper instructor's involvement and it could be more time and resource consuming. In such educational settings instead of giving collective lectures the instructor should provide a personal or group guidance.

Adaptivity and adaptability are inseparable from personalized learning. Adaptation in e-learning could be seen as a method to create a learning experience for the learner but also for the instructor. In order to increase the performance of pre-defined criteria (like economic, educational, user satisfaction-based or time-based) instructor must configure a set of specific elements (usually based on content, interface, order, time, assessment, and so on).

Three essential inputs exist in a balanced formula for adaptation: the user (learner, student), the teacher (tutor, instructor), and the set of pre-defined rules made by the learning instructor i.e. designer.

Usually, three essential types of adaptation have been proposed in literature:

1. **Interface-based (also known as adaptive navigation)**. It relates to elements and options of the interface and usability and adaptability: where particular elements are positioned on the screen, which properties are defined (size, colour, etc.) and so on.
2. **Learning flow-based**. The learning process is dynamically adapted to the sequence in appropriate (different) ways the contents of the course is delivered.
3. **Content-based**. In such systems resources and activities dynamically change their actual content (for example systems based on adaptive presentation).

Also, there are some key researchers in e-learning area (Brusilovsky 2004) who recognized and proposed several additional kinds of adaptation:

1. **Interactive problem solving support**. In order to get an appropriate solution to a problem the learner is guided (from an online or offline tutor or from a predefined set of rules) to the next step in the learning process.
2. **Adaptive information filtering**. In order to provide relevant and categorized outputs to the learner system takes care of appropriate information retrieval.
3. **Adaptive user grouping**. Such kind of systems allows ad hoc creation of learners' groups and collaborative support for performing particular tasks.

In (Burgos 2011) authors proposed further extension of the classification:

1. **Adaptive evaluation**. Based on the performance of the learner and the guidance of the tutor, elements like the actual content, the evaluation model, and the running of a test can be changed.
2. **Changes on-the-fly**. In these systems there is the possibility to adapt/modify a course on-the-fly by the instructor in run-time.

Adaptation and personalization are posing new research and development challenges to modern e-learning systems. E-learning definitely becomes smarter from the exploration of the efficiency of interaction analysis methods that empower these systems with adaptation and personalization. Recently advanced Artificial Intelligence techniques have been exploited for implementation of smarter online (but also blended learning) scenarios, including complex character of collaboration.

Analysis of trends in modern e-learning systems showed that the most popular types of personalization in today's e-learning systems are (Klašnja-Milićević et al. 2011):

- *Learning style identification*. Personalization of the system is based on the identified learning styles of each user of the system.
- *Recommendation systems*. These systems are used to recommend appropriate educational material to the learner and to select optimal paths through the learning materials.
- *Link adaptation*. The system modifies the appearance and/or availability of every link that appears on a course Web page, in order to show the learner, whether the link leads to interesting new information, to new information the

learner is not ready for, or to a page that provides no new knowledge. The system makes some links inaccessible to the learner if the system estimates from the learner model that such links take him/her for the irrelevant information.

- *Personalised pedagogical agents*. The demand of modern e-learning systems is to make learning process more challenging, exciting and highly interactive. Usually e-learning environments are equipped with different kinds of agents that support more intelligent and human-like (teacher-to-student) communication within the system. Personal, pedagogical avatars (Haake and Gulz 2008) are a way to facilitate higher quality of delivering topics and assessing acquired knowledge.

References

Baguley, M., Danaher, P. A., Davies, A., De George-Walker, L., Matthews, K. J., Midgley, W., et al. (2014). *Educational learning and development: building and enhancing capacity*. Palgrave Macmillan.

Brusilovsky, P. (2004). KnowledgeTree: a distributed architecture for adaptive e-learning. In *WWW Alt. '04: Proceedings of the 13th international World Wide Web conference on Alternate track papers & posters* (pp. 104–113). http://doi.org/10.1145/1013367.1013386

Burgos, J. L. M. (2011). Semantic web standards. *SNET Computer Engineering*. Retrieved from http://www.pdffiller.com/948565-semantic-web-standards_burgos-Semantic-Web-Standards—SNET-Various-Fillable-Forms-snet-tu-berlin

Haake, M., & Gulz, A. (2008). Visual stereotypes and virtual pedagogical agents. *Educational Technology and Society, 11*(4), 1–15.

Khemaja, M., & Taamaallah, A. (2016). Towards situation driven mobile tutoring system for learning languages and communication skills: Application to users with specific needs. *Journal of Educational Technology & Society, 19*(1).

Klašnja-Milićević, A., Vesin, B., Ivanović, M., & Budimac, Z. (2011). E-Learning personalization based on hybrid recommendation strategy and learning style identification. *Computers & Education, 56*(3), 885–899.

Chapter 3
Personalization Based on Learning Styles

Abstract It is obvious that different learners have different preferences, needs and approaches to learning. Psychologists distinguish these differences as individual learning styles. Therefore, it is very important to accommodate for the different styles of learners through learning environments that they prefer and find more efficient. Learning styles can be defined as unique manners in which learners begin to concentrate on, process, absorb, and retain new and difficult information. While there are still many open issues with respect to learning styles, the learning style models agree that learners have different ways in which they prefer to learn. This chapter presents the bases of electronic learning techniques for personalization of learning process based on individual learning styles and the possibilities of their integration in e-learning systems.

It is obvious that different learners have different preferences, needs and approaches to learning. Psychologists call these differences as the individual learning styles. Therefore, it is very important to accommodate for the different styles of learners through learning environments that they prefer and find more efficient. Learning styles can be defined as unique manners in which learners begin to concentrate on, process, absorb, and retain new and difficult information (Dunn et al. 1984). They are distinctive individual patterns of learning, which vary from person to person. It is necessary to determine what is most likely to trigger each learner's concentration, how to maintain it, and how to respond to his or her natural processing style to produce long term memory and retention (Dorça et al. 2016).

The term *learning styles* refers to the concept that individuals differ in regard to what mode of instruction or study is the most effective for them (Pashler et al. 2009). Proponents of learning style assessment contend that the optimal instruction requires diagnosing individual learning styles and tailoring instruction accordingly. Many learning style models exist in literature, e.g. the learning style model by Felder and Silverman (1988), Kolb (1984), Mumford and Honey (1986), Pask (1976). While there are still many open issues with respect to learning styles, the learning style models agree that learners have different ways in which they prefer to learn.

© Springer International Publishing Switzerland 2017

A. Klašnja-Milićević et al., *E-Learning Systems*,

Intelligent Systems Reference Library 112, DOI 10.1007/978-3-319-41163-7_3

3.1 Learning Style's Theories

According to Felder and Silverman (1988), a learning style model classifies learners according to the ways in which they receive and process information and acquire knowledge, while a teaching style model classifies teaching methods according to how well they address the proposed learning style components. Although the existence of one unique learning style model and assessment tool, which will generally be accepted and applied, could brought the benefits to all stakeholders, today in the field of the learning style theories there is the opposite situation. Namely, since the early 60s to these days researchers have built over 70 different learning style models accompanied by the instruments for their assessment. They mostly differ in a way in which they distinguish the leading factors of the learning process.

All learning style models can be classified into five learning style families (Coffield et al. 2004; Dorça et al. 2016; Halawa et al. 2015), according to the cognition of the learning style concept:

- Learning styles and preferences are **constitutionally based**, including four modalities: visual, auditory, kinaesthetic and tactile;
- Learning styles reflect **features of the cognitive structure**, including patterns of ability;
- Learning styles are one component of **personality type**;
- Learning styles are **flexibly stable learning preferences**;
- Move on from learning styles to **learning approaches, strategies, orientations and conceptions**.

Figure 3.1 shows the list consisting of the most frequently used learning style models, classified within the learning styles families.

The most influential models and instruments of learning styles are chronologically described below.

Witkin 1962—Group Embedded Figures Test (GEFT). Witkin distinguishes field-independent from field-dependent learners (Witkin et al. 1975). The main characteristics of field-independent learners are that they work with an internal frame of reference, with an intrinsic motivation and self-directed goals; they are less affected by criticism and define their own learning strategy. On the opposite, field-dependent learners are operating using an external frame of reference and extrinsic motivation, so they need structuring and assistance from the instructor. They have also a need to interact with other learners.

Kolb 1976—Learning Style Inventory (LSI). Kolb considers learning styles as both flexible and stable, outlining that they are not fixed characteristic of the personality, but are relatively stable patterns of learners' behavior. Kolb's learning style model is based on the theory of experiential learning which incorporates growth and development the personality. Learning Style Inventory is an assessment tool improved through the long period of almost 40 years of using and testing its validity.

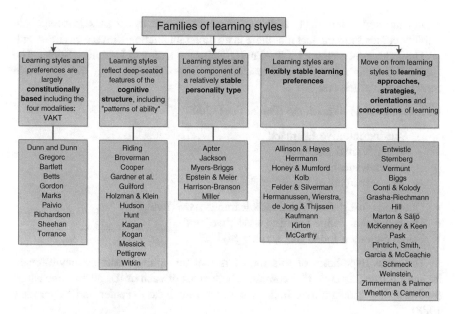

Fig. 3.1   Families of learning styles (Coffield et al. 2004)

Gregorc 1977—Gregorc Mind Styles Delineator (MSD). According to Gregorc's model, there are two dimensions of learners' unconscious abilities of perception and ordering, determining his learning style: concrete-abstract and sequential-random. Individuals may be strong in one or two of the following categories: concrete sequential, concrete random, abstract sequential and abstract random.

Dunn and Dunn 1979—Learning Style Questionnaire (LSQ). The main characteristic of this model is its user-friendly style and its orientation on examination of motivational factors, social interaction, physiological elements and environmental factors. Learning Style Questionnaire examines learners' preferences for 22 different factors. High learners' preferences are the initial sign for teachers to make changes in learning environment by changing the teaching process (through introduction of different sound, design, teaching time, classroom light, mobility or adoption of a new kind of teaching techniques).

Curry 1983—Onion model. In his model, using the onion metaphor (Curry 1983) grouped different approaches into three main types, depicting them as different layers of an onion. He named the outmost layer as 'instructional preferences' as a layer that interact the most with the learning environment and the one to be the least stable but less important for learning. The second level, the middle one, is called 'information processing style', conceived of as the individual's intellectual approach to assimilating information, more stable than instructional preferences, but still modifiable by learning strategies. Finally, the third and innermost layer is named as 'cognitive personality style', the most stable and more significant in complex learning. The stability comes from the fact that there is no direct interaction with the environment.

The Felder-Silverman 1988—Felder-Solomon Index of Learning Styles. The Felder-Silverman's learning and teaching style model consists of the four basic components, which are:

- Learning style dimensions (Felder et al. 2000):

  - Sensing/Intuitive Learners;
  - Visual/Verbal Learners;
  - Active/Reflective Learners;
  - Sequential/Global Learners;

- Questions that define a student's learning style (five),
- Questions that define teaching style (five) and
- Felder-Solomon Index of Learning Styles.

The main hypothesis of this model is that an optimal learning environment includes a teaching style that combines both sides of each of the given dimensions (both sensing and intuitive, both visual and verbal etc.) (Felder and Silvermans 1988).

Riding 1991—Cognitive Styles Analysis (CSA). According to Riding's model, learners may improve their learning strategies through the learning process. He outlines two different dimensions of the learning style: holist-analytic, focused on the ways of organizing information and verbaliser-imager, considering the ways of representing information. It is important to mention that these two dimensions are independent on learners' intelligence.

Herrmann Brain 1995—Dominance Instrument (HBDI). Herrmann Brain's model considers learning styles as learned patterns of student's behaviour, including development and person's creativity as important learning factors. This is contrary to the theories of learning styles as fixed personality features.

Vermunt 1996—Inventory of Learning Styles (ILS). This model is based on the idea of integration of cognitive, affective, meta-cognitive and co-native processes. It examines different elements of the learning process like learning strategies, motivation for learning and preferences for organizing information. The model is based on interviews, primarily for application among university students and teachers, but its extended version has application in business environment and among younger learners.

Apter 1998—Motivational Style Profile (MSP). The Apter's theory considers human behaviour and experience, not as fixed personality types, but as a dynamic interplay between 'reversing' motivational states. According to Apter's model, there are four domains of experience (means-ends, rules, transactions and relationships) in which we can find an interaction between emotion, cognition and volition.

Sternberg 1998—Thinking Styles. The Sternberg's model of learning styles is based on a new theory of mental self-government. He proposes 13 thinking styles, which are based on the following concepts: functions, forms, levels, scopes and meanings of government.

Jackson 2002—Learning Styles Profiler (LSP). Jackson's model describes four learning styles named Initiator, Analyst, Reasoner and Implementer. The model is designed for use both in education and business.

3.2 Learning Styles in E-Learning Systems

Researchers in the field of education mainly share the opinion about the importance of learning style in learning process in general, and especially in online learning (Watkins 2015). Many of them focus on finding the evidence of its positive influence on the performance and the results of the learning process. The installation of knowledge about learning styles in an online learning system as a tool for facilitating of learning and its personalization has been emphasized in a lot of studies. There are numerous studies dealing with the issue of students' behaviour in online learning environments, and its connection to the corresponding learning styles (Cristea and Stash 2006; Dorça et al. 2016; Graf et al. 2010; Liegle and Janicki 2006; Lu et al. 2007).

There are still different opinions among researchers about the causal relationship between the adaptability to learning style of the learner in an online learning environment, and the quality of the learning process. Most of the researchers consider it important for an adaptive learning system to have a mechanism for accommodating to a specific learning style of a certain learner (Coffield et al. 2004; Popescu 2010).

On the other hand, there are some studies which minimize the importance of adaptation to learning styles in e-learning systems (Freedman and Stumpf 1980; Holodnaya 2002). The results of these studies mainly fail to show any significant enhancement in the online learning process arising from the adaptability of the e-learning system to the learning style. Furthermore, in psychological research conducted by Holodnaya (2002), learning in an environment which is incompatible with learner's preferences could encourage learner in developing some new skills.

However, positive opinions on the influence of the adaptability of an e-learning environment to a specific learning style during online learning process prevail. There are a lot of studies where this attitude is substantiated by experiments with measurable results.

Popescu et al. (2007) consider that e-learning systems with flexibility of changing instructions through the learning process with respect to the learner's learning style preferences could increase efficiency, effectiveness and satisfaction of learner.

According to Stash et al. (2006), it is important to incorporate a knowledge about learner's learning style in e-learning environment. Such e-learning environment, which provides the learners with a possibility of choosing the most suitable way of learning could enhance its results.

A group of studies emphasizes a navigational behaviour as one of the learner's characteristics that is important for more accurate determination of learning style in online learning systems. That must be taken into account when creating an adaptive learning system which aims to achieve an acceptable personalization function.

An adaptive navigation support of a learning management system (its ability to recommend learners the most appropriate learning steps and personalized way of passing through the learning material) is marked as one of the leading technologies for achieving adaptability of online learning systems (Brusilovsky 2004). Numerous studies focus on the navigational behaviour of learner considering learning styles in learning processes. The most of the existing LMS, like CS383 (Carver et al. 1999), WELSA (Popescu 2010), and TSAL (Hwang et al. 2008), have the adaptability features based on the learner's navigational behaviour considering learning styles in learning processes which are not obligatory online. On the other hand, Graf et al. (2007) focused their research on students' behaviour in online environment only. They have conducted a study about the navigational behaviour of students in an online course within a learning management system, investigating the causality between students' different learning styles, and their activities and preferences during online learning process. The results of the study can be summarized as follows:

- Students' different learning styles cause the usage of different strategies for learning and navigating through the course.
- Monitoring student's navigating behaviour and processing the output data could be useful for enhancing the adaptability of the learning management system.
- The information about differences in student's navigational behaviour can be used for developing a new pattern in student modelling that identifies learning styles automatically from students' behaviour in an online course. This can improve former student modelling approaches, based mainly on counting the students' visits of learning objects, measuring students' time spent on different kinds of learning objects or assessing their performance of different parts of learning courses (exercises, tests, etc.).

Previous conclusions were implemented in the development of LMS called DeLeS, in which Kumar, (Kinshuk et al. 2011) introduced the extended way for automatic identification of learning styles in an online learning system. In DeLeS there are two kinds of patterns for identification of learning styles:

1. Behaviour patterns—for detecting learning styles, and
2. Additional sources—for improving the identification of learning styles.

So, the architecture of DeLeS consists of two basic components:

1. Extraction component (for extracting data from the learning system's database), and
2. Calculation component (for quantification of learning styles from students' behaviour).

This extended architecture of DeLeS includes three kinds of data sources: data from behaviour patterns, navigation patterns and cognitive abilities. The results of this study showed that incorporation of data from extended sources about the students' learning behaviour (navigation patterns and cognitive traits), has its justification in improving the accuracy of the learning style identification.

The importance and the influence of knowledge about learning styles in an adaptive learning environment is doubtlessly worthy of further research. Irrespective of different opinions in relevant studies, we strongly support the strand of research in which the modelling of learning style in LMS has to be a built-in functionality. This could provide the learners with a possibility to define their learning style, and to choose from different instructional strategies offered.

3.3 Learning Style Index by Felder and Soloman

According to the comprehensive study of the e-learning environment, we selected Felder and Soloman's data collection instrument, called Index of Learning Styles (ILS) (Soloman and Felder 2005). The ILS is a 44-questions, freely available, multiple-choice learning style instrument, which assesses variations in individual learning style preferences across four dimensions or domains. These are *Information Processing*, *Information Perception*, *Information Reception*, and *Information Understanding*. Within each of the four domains of the ILS there are two categories (see Table 3.1):

Table 3.1 Characteristics of ILS based on Soloman and Felder (2005)

Active	Reflective
Work in groups	Work alone
Preference to try out new material immediately (ask, discuss, and explain)	Preference to take time to think about a problem
Practical (experimentalists)	Fundamental (theoreticians)
Sensing	Intuitive
More patient with details	More interested in overviews and a broad knowledge (bored with details)
By standard methods	Innovations
Senses, facts and experimentation	Perception, principles and theories
Visual	Verbal
Preference to perceive materials as pictures, diagrams and flow chart	Preference to perceive materials as text
Global	Sequential
Prefer to get the big picture first	Prefer to process information sequentially
Assimilate and understand information in a linear and incremental step, but lack a grasp of the big picture	Absorb information in unconnected chunks, and achieve understanding in large holistic jumps without knowing the details

- **Information Processing**: *Active* and *Reflective* learners,
- **Information Perception**: *Sensing* and *Intuitive* learners,
- **Information Reception**: *Visual* and *Verbal* learners,
- **Information Understanding**: *Sequential* and *Global* learners.

The preferred learning style can be investigated by offering the learner a free choice between several different forms of examples, activities or explanations at first, and by observing behaviour and pattern in which (s)he makes the choices.

3.3.1 Information Processing: Active and Reflective Learners

Within *Information Processing* domain, we can distinguish example-oriented learners, called *Reflectors*, and activity-oriented learners, called *Activists* (Kolb 1984). Active learners tend to retain and understand information best by doing something active with it—discussing or applying it or explaining it to others. Reflectors are people who tend to collect and analyse data before taking an action. They may be more interested in reviewing other learners' and professional opinions than doing real activities. In the e-learning systems, a learner with the active learning style can be presented with an activity first, then an example, explanation and theory. For the learner with the reflective style this order would be different— (s)he is shown an example first, then an explanation and theory, and finally (s)he is asked to perform an activity.

3.3.2 Information Perception: Sensing and Intuitive Learners

Within *Information Perception* domain, sensing learners, called *Sensors*, tend to be patient with details and good at memorizing facts and doing hands-on (laboratory) work. On the other hand, intuitive learners, called *Intuitors*, may be better at grasping new concepts and are often more comfortable with abstractions and mathematical formulations than sensing learners. *Sensors* often prefer solving problems using well-established methods, and dislike complications and surprises. On the other hand, *Intuitors* like innovation and dislike repetition.

Sensors tend to be more practical and careful than *Intuitors*. *Intuitors* tend to work faster and to be more innovative than *Sensors*. For example, it is assumed that sensing learners will be interested in additional materials; therefore this kind of material can be recommended to them. *Intuitors* are provided with abstract material, formulas and concepts. Adequate explanations in e-learning system can be presented to them in the form of block diagrams or exact syntax rules.

3.3.3 Information Reception: Visual and Verbal Learners

Within *Information Reception* domain *Visual* learners remember best what they see—pictures, diagrams, flow charts, time lines, and demonstrations (Klašnja-Milićević et al. 2011). *Verbal* learners get more out of words—written and spoken explanations.

3.3.4 Information Understanding: Sequential and Global Learners

Within *Information Understanding* domain *Sequential* learners tend to follow logical stepwise paths in finding solutions. On the other hand, *Global* learners may be able to solve complex problems quickly or put things together in novel ways once they have grasped the big picture, but they may have difficulty explaining how they did it. *Sequential* learners prefer to go through the course step by step, in a linear way with each step following logically from the previous one, while global learners tend to learn in large leaps, sometimes skipping learning objects and jumping to more complex material. According to these characteristics of *Sequential* learning style, these learners are led through educational material by a predefined order. On the other hand, *Global* learners are provided with an overall view of the course, with short explanations of each unit and options for accessing the unit they are interested in by clicking the unit hyperlinks rather than following sequential order.

References

Brusilovsky, P. (2004). KnowledgeTree: A distributed architecture for adaptive e-learning. In *WWW Alt. '04: Proceedings of the 13th International World Wide Web Conference on Alternate Track Papers & Posters* (pp. 104–113). http://doi.org/10.1145/1013367.1013386

Carver, C. A., Howard, R. A., & Lane, W. D. (1999). Addressing different learning styles through course hypermedia. *IEEE Transactions on Education, 42*(1), 33–38.

Coffield, F., Moseley, D., Hall, E., & Ecclestone, K. (2004). *Learning styles and pedagogy in post 16 learning: A systematic and critical review.* The Learning and Skills Research Centre.

Cristea, A., & Stash, N. (2006). AWELS: Adaptive web-based education and learning styles. In *Sixth IEEE International Conference on Advanced Learning Technologies (ICALT'06)* (pp. 1135–1136). http://doi.org/10.1109/ICALT.2006.1652660

Curry, L. (1983). An organization of learning styles theory and constructs.

Dorça, F., Araújo, R., de Carvalho, V., Resende, D., & Cattelan, R. (2016). An automatic and dynamic approach for personalized recommendation of learning objects considering students learning styles: An experimental analysis. In *Informatics in education* (Vol. 15, pp. 45–62). Vilnius University.

Dunn, R., Dunn, K., & Freeley, M. E. (1984). Practical applications of the research: Responding to students' learning styles–step one. *Illinois State Research and Development Journal, 21*(1), 1–21.

Felder, R., & Silverman, L. (1988). Learning and teaching styles in engineering education. *Engineering Education, 78*, 674–681. http://doi.org/10.1109/FIE.2008.4720326

Felder, R., Silverman, L., & Solomon, B. (2000). Index of learning styles (ILS). *Skynet.ie.*

Freedman, R. D., & Stumpf, S. A. (1980). Learning style theory: Less than meets the eye. *Academy of Management Review, 5*(3), 445–447.

Graf, S., Liu, T. C., & Kinshuk. (2010). Analysis of learners' navigational behaviour and their learning styles in an online course. *Journal of Computer Assisted Learning, 26*(2), 116–131. http://doi.org/10.1111/j.1365-2729.2009.00336.x

Graf, S., Viola, S. R., & Leo, T. (2007). In-depth analysis of the Felder-Silverman learning style dimensions. *Journal of Research on Technology in Education, 40*, 79–93. http://doi.org/10. 1080/15391523.2007.10782498

Halawa, M. S., Hamed, E. M. R., & Shehab, M. E. (2015). Personalized E-learning recommendation model based on psychological type and learning style models. In *2015 IEEE Seventh International Conference on Intelligent Computing and Information Systems (ICICIS)* (pp. 578–584).

Holodnaya, M. A. (2002). *Psychology of intelligence: The paradoxes of the study.* Petersburg: Peter.

Hwang, G.-J., Tsai, P.-S., Tsai, C.-C., & Tseng, J. C. R. (2008). A novel approach for assisting teachers in analyzing student web-searching behaviors. *Computers & Education, 51*(2), 926–938.

Kinshuk, K., Chang, M., Dron, J., Graf, S., Kumar, V., Lin, O., ... Yang, G. (2011). Transition from e-learning to u-learning: Innovations and personalization issues. In *2011 IEEE International Conference on Technology for Education (T4E)* (pp. 26–31).

Klašnja-Milićević, A., Vesin, B., Ivanović, M., & Budimac, Z. (2011). E-learning personalization based on hybrid recommendation strategy and learning style identification. *Computers & Education, 56*(3), 885–899.

Kolb, D. (1984). *Individuality in learning and the concept of learning styles* (pp. 61–98). Englewood Cliffs, New Jersey: Prentice Hall.

Liegle, J. O., & Janicki, T. N. (2006). The effect of learning styles on the navigation needs of web-based learners. *Computers in Human Behavior, 22*(5), 885–898. http://doi.org/10.1016/j. chb.2004.03.024

Lu, H., Jia, L., Gong, S., & Clark, B. (2007). The relationship of kolb learning styles, online learning behaviors and learning outcomes Shu-hong Gong. *Learning, 10*, 187–196.

Mumford, A., & Honey, P. (1986). *The manual of learning styles.* Maidenhead, Berkshire: P. Honey, Ardingly House.

Pashler, H., McDaniel, M., Rohrer, D., & Bjork, R. (2009). Learning styles concepts and evidence. *Psychological Science in the Public Interest, Supplement, 9*, 105–119. http://doi.org/10.1111/j. 1539-6053.2009.01038.x

Pask, G. (1976). Styles and strategies of learning. *British Journal of Educational Psychology, 46* (2), 128–148.

Popescu, E. (2010). Adaptation provisioning with respect to learning styles in a web-based educational system: An experimental study. *Journal of Computer Assisted learning, 26*(4), 243–257.

Popescu, E., Bădică, C., & Trigano, P. (2007). Rules for learner modeling and adaptation provisioning in an educational hypermedia system. In *Proceedings—9th International Symposium on Symbolic and Numeric Algorithms for Scientific Computing, SYNASC 2007* (pp. 492–499). http://doi.org/10.1109/SYNASC.2007.72

Soloman, B. A., & Felder, R. M. (2005). Index of learning styles questionnaire. *NC State University.* Available Online at: http://www.Engr.Ncsu.Edu/learningstyles/ilsweb.Html. Last Visited on May 14, 2010.

Stash, N., Cristea, A. I., & De Bra, P. (2006). Adaptation to learning styles in e-learning: Approach evaluation.

Watkins, C. (2015). Meta-learning in classrooms. In *The SAGE handbook of learning* (p. 321).

Witkin, H. A., Moore, C. A., Goodenough, D. R., & Cox, P. W. (1975). Field-dependent and field-independent cognitive styles and their educational implications. *ETS Research Bulletin Series, 1975*(2), 1–64.

Chapter 4
Adaptation in E-Learning Environments

Abstract In e-learning systems, learners usually does not visit educational materials linearly, but have access to different materials via a number of links to other lessons or teaching units. In modern Web-based learning environments, the authors avoid creation of static learning material that is presented to the learner in a linear way, due to the large amount of interdependences and conditional links between the various pages. Often, authors create multiple versions of learning resources so the system can propose to the learner the appropriate one. This leads to the learning concept known as content adaptation. The order of visiting educational material can be influenced by manipulating the hypertext links. This process is called link adaptation. The most popular content and link adaptation techniques used in e-learning environments are presented in this chapter. The chapter also covers basic principles of adaptive educational hypermedia systems.

In e-learning systems, learners usually do not visit educational materials linearly, but have access to different materials with the assistance of a number of links to other lessons or teaching units. These links can be in the form of navigational elements (buttons) or in the form of specially emphasized piece of text in the form of hyperlinks.

In modern Web-based learning, the authors avoid creation of static learning material that is presented to learner in a linear way, due to the large amount of interdependence and conditional links between the various pages (De Bra et al. 2003). Therefore, methods and techniques of adaptive hypermedia provide information to learners, whether some links:

- lead to material that learner is not ready for,
- propose visit to some useful sites or
- provide additional explanations.

Adaptive educational hypermedia systems use data from the learner model in order to adapt content and links of hypermedia course according to needs of each learner (Henze et al. 2004). These systems provide personalized learning using technology that continuously measures learner's knowledge and progress.

© Springer International Publishing Switzerland 2017
A. Klašnja-Milićević et al., *E-Learning Systems*,
Intelligent Systems Reference Library 112, DOI 10.1007/978-3-319-41163-7_4

Collected data on learner are used to further customize the display of educational materials to the needs, pace of work, desires and goals of the learner.

4.1 Adaptive Educational Hypermedia

Adaptive hypermedia is an alternative to the traditional "one-size-fits-all" approach in the development of hypermedia systems. Adaptive hypermedia systems build a model of the goals, preferences and knowledge of each individual user. This model has been used in interaction between system and user, in order to adapt information and knowledge to the needs of that user. For example, to a learner in an adaptive educational hypermedia system will be presented educational material that is adapted specifically to his knowledge of the subject (De Bra 1999), and a suggested set of the most relevant links to proceed further (Brusilovsky et al. 1996).

Adaptive *Hypermedia systems* are a combination of text, hypertext, hyperlinks, audio and video materials in order to form the non-linear medium of information. *Adaptive Hypermedia* is a technique of adjustment (presentation, highlighting or concealment) of hyperlinks with the aim of selecting the appropriate content to the user. There are two main reasons for the use of adaptive hypermedia: avoiding the problems that occur when the material is read by the planned schedule and better adaptation to individual differences among learners (De Bra 2006).

Adaptive Hypermedia systems can sort, highlight or hide the links to individual Web pages in order to select content to be shown and generate recommendations to learners, which resources can visit based on analysis of their personal characteristics (objectives, needs and level of knowledge). Learners can also receive information about the importance and significance of certain elements of the educational materials.

Adaptive Hypermedia systems combine hypermedia with techniques for user modelling and can be used in various application fields, dominated by the education (De Bra 2006). The aim of adaptive hypermedia is to overcome the problems of the presentation of the same content to different learners. An approach using adaptive hypermedia involves collecting data about each learner and adaptation of learner-system interaction on the basis of this information.

Important question arise when speaking about any kind of adaptive systems: What can be adapted in this system? Which features of the system can differ for different users? What is the space of possible adaptations?

In adaptive hypermedia, the adaptable space is quite limited: there are not so many features which can be altered. At some level of generalization, hypermedia consists of a set of nodes or hyper documents (for the purpose of brevity, we will call them "pages") connected by links. Each page contains some local information and a number of links to related pages. Hypermedia systems can also include an index and a global map which provides links to all accessible pages. Adaptive hypermedia offers two general categories of adaptation:

- **content-level** adaptation: the content of regular pages can be adapted and
- **link-level** adaptation: the links that appear on a course Web page can be the links from regular pages, index pages, and maps.

Therefore, it can be distinguished content-level and link-level adaptation as two different classes of hypermedia adaptation. First one is known as adaptive presentation (content adaptation) and the second one is known as adaptive navigation support (link adaptation):

- **Content adaptation**—the system supports presentation of the content in different ways, according to the domain model (concepts, their relations, prerequisites for presentation of material, etc.) and information from the learner model (De Bra 2006).
- **Link adaptation**—the system modifies the appearance and/or availability of every link that appears on a course Web page, in order to show the learner, whether the link leads to interesting new information, to new information the learner is not ready for, or to a page that provides no new knowledge (Romero et al. 2007). The system can make some links inaccessible to the learner if the system estimates from the learner model that such links take him/her for the irrelevant information. The system may assume that less successful learners will be interested in additional material. Therefore, those learners may click the link for additional material on the interface.

4.2 Content Adaptation

E-learning systems provide additional explanations using so-called conditional pages—links are displayed, highlighted or removed as needed, based on the information from the appropriate learner model. With this model, the system can determine whether certain resources correspond to the identified learner profile and they are designed to present the learner with important information.

Usually in every course i.e. educational material there are a lot of cross-references between different chapters/sections. In a paper textbook the author knows whether such a reference is a forward or a backward reference, and thus whether the reference is to a concept 'not yet known' as opposed to 'already known'. In an on-line course with free navigation through text the author cannot know whether a reference is a forward or backward reference. However, it isn't difficult to track a learner's path through the course text, and thus for the system to know whether for this learner at this time a reference is a forward or backward reference. So if an author creates these two versions of the reference the system can choose and present the appropriate one. This leads to the first important form of adaptation—*content adaptation* (Brusilovsky 1998).

There are basically three cases where content adaptation is appreciated. (Brusilovsky 1998):

- When a reference is made to a concept the learner does not yet know. A short prerequisite explanation can be inserted to prepare the learner for the rest of the description of the topic.
- Sometimes the current concept can be elaborated upon further in case the related concept is already known, or when the knowledge level of the learner is already high. For these "expert" users an additional explanation can be given that is beyond the level of the average learner (at the time of visiting the current page).
- Sometimes an interesting comparison is possible with another concept, but only if that concept is already known. Such a comparative explanation between the concepts can automatically be shown on the page of the second concept studied by the learner.

When educational material is carefully and professionally prepared in content adaptation manner the learner may easily use it and even not be aware that content adaptation is being performed. If content adaptation is frequently performed in an educational material the learners may become confused, thinking that they are "missing out" on some of the information in the course.

4.3 Link Adaptation

The differences in learners' knowledge, caused by choosing different paths through the educational material can be compensated for to some extent, but sometimes it is necessary to guide the learners in a certain direction or to keep learners away from some learning material they are really not ready for (meaning that a short prereq-uisite explanation is not sufficient to prepare the learner for the rest of the description of the topic). The order of visiting educational material can be influ-enced by manipulating the hypertext links. This reveals the second form of adap-tation: link adaptation (Brusilovsky 1998).

The basic idea with link-adaptation is to change or annotate the link structure in such a way that the user is guided towards interesting, relevant information, and kept away from non-relevant information. Link-adaptation tries to make simpler the link structure to reduce orientation problems, while retaining a lot of navigational independence, a typical feature of hypermedia systems.

The techniques found in (De Bra et al. 2000; Kobsa et al. 2001) for link-adaptation are:

- direct guidance (e.g. a "next" button);
- link sorting (like in search engines);
- link hiding (hide non-relevant links, but keep anchor text);
- link annotation (e.g. use colours to indicate relevance);
- link disabling (make non-relevant links not work);
- link removal (remove non-relevant link anchors);
- map adaptation (provide a personalized overview).

The system modifies the appearance and/or availability of every link that appears on a course Web page, in order to show the learner whether the link leads to interesting new information, to new information the learner is not ready for, or to a page that provides no new knowledge (Romero et al. 2007). The system can make some links inaccessible to the learner if the system estimates from the learner model that such links take him/her for the irrelevant information. The system may assume that less successful learners will be interested in additional material. Therefore, those learners using system interface may click the link for additional material.

Link adaptation can be done in several ways, because the links have a position in the page determined by a link anchor and a destination:

- *Adaptive link sorting* means that the position of links has an influence on the learner's behaviour. This foreseen meaning can be used to sort lists of links so that the most appropriate links for the current learner, in his current state of mind, are placed at the top. Links can also be made "available" through a system of menus and submenus. A usual technique is to show a list of chapters, and the list of sections of the "current" chapter as list of links. By having only the sections of one chapter listed at the time the learner is suggested to select only sections from that chapter and not switching between sections in different chapters.
- Links are presented through a *link anchor*. Link anchor can be a word, a phrase or an icon, displayed in a clear way as a link. On the Web, links are typically represented by anchors that appear in blue and that are underlined. Images or icons that are link anchors typically have a blue border. There are different ways in which the learner can be guided towards or away from certain links by changing the presentation of the link anchor. Most Web browsers change the colour of the link anchor after the link is visited (usually become purple). Nevertheless, it is possible to change the appearance of link anchors in other ways, and it is also possible to add icons to indicate a special meaning of a link. The AHA! System (De Bra et al. 2005) has a default link adaptation technique of *link hiding*. Link can be inaccessible or invisible to the learner if the system finds out that it leads to irrelevant information. Colour of link anchors can be blue, purple or black, and link anchors are not underlined in AHA! so the black links are effectively hidden. They look just like normal text. It is also possible to change this colour scheme and use three visible colours, resulting in a technique of link annotation. Other systems, like ELM-ART (Weber and Brusilovsky 2001) and Interbook (Brusilovsky et al. 2004) for instance, use icons to recommend 'for' or 'in contradiction of' certain links (green and red balls).
- Learners expect a link anchor to always correspond to the same link, and hence to lead to the same link destination. On the other hand, when pages are generated dynamically (e.g. by an adaptive system) there is no technical reason why the same anchor cannot lead to a different destination, depending on specific conditions. A link can for instance lead to a shortened description of a topic or a detailed description depending on the learner's progress. In an educational environment, in (De Bra 2006), the use of an adaptive course (with adaptive

tests) is based on the outcome of a test. Result of the test would automatically lead the learner to a particular chapter at a beginner's, intermediate or advanced level.

There is no adaptive hypermedia system that supports all the methods and techniques presented in this section. Using all link-adaptation techniques simultaneously would lead to a confused and non-functional system.

References

Brusilovsky, P. (1998). Adaptive educational systems on the World-Wide-Web : A review of available technologies. In *Proceedings of Workshop "WWW-Based Tutoring" at 4th International Conference on Intelligent Tutoring Systems (ITS'98), San Antonio, TX*.

Brusilovsky, P., Karagiannidis, C., & Sampson, D. (2004). Layered evaluation of adaptive learning systems. *International Journal of Continuing Engineering Education and Life Long Learning, 14*(4), 402–421.

Brusilovsky, P., Schwarz, E., & Weber, G. (1996). ELM-ART: An intelligent tutoring system on World Wide Web. In *Intelligent tutoring systems* (pp. 261–269).

De Bra, P. (1999). Design issues in adaptive web-site development. In *Proceedings of the 2nd Workshop on Adaptive Systems and User Modeling on the WWW* (pp. 29–39).

De Bra, P. (2006). Web-based educational hypermedia. *Data Mining in E-Learning (Advances in Management Information), 4*, 3–16.

De Bra, P., Brusilovsky, P., & Houben, G.-J. (2000). Adaptive hypermedia—From systems to framework. *Journal ACM Computing Surveys, 31*(4), 1–6. http://doi.org/10.1145/345966. 345996.

De Bra, P., Stash, N., & Smits, D. (2005). Creating adaptive web-based applications. In: *Tutorial at the 10th International Conference on User Modeling*, 1–33.

De Bra, P., Aerts, A., Berden, B., De Lange, B., Rousseau, B., Santic, T. & Stash, N. (2003). AHA! The adaptive hypermedia architecture. In *Proceedings of the fourteenth ACM conference on Hypertext and hypermedia* (pp. 81–84). ACM.

Henze, N., Dolog, P., & Nejdl, W. (2004). Reasoning and ontologies for personalized e-learning in the semantic web. *Educational Technology & Society, 7*(4), 82–97.

Kobsa, A., Koenemann, J., & Pohl, W. (2001). Personalised hypermedia presentation techniques for improving online customer relationships. *The Knowledge Engineering Review, 16*(02), 111. http://doi.org/10.1017/S0269888901000108.

Romero, C., Ventura, S., Delgado, J. A., & De Bra, P. (2007). Personalized links recommendation based on data mining in adaptive educational hypermedia systems. In *Creating New Learning Experiences on a Global Scale* (pp. 292–306). Berlin: Springer.

Weber, G., & Brusilovsky, P. (2001). ELM-ART : An adaptive versatile system for web-based instruction. *International Journal of Artificial Intelligence in Education, 12*(4), 351–384. http://doi.org/10.1.1.66.6245.

Chapter 5
Agents in E-Learning Environments

Abstract A recent trend in the field of e-learning and tutoring systems is to utilize agent technology and develop and use different kinds of agents in virtual learning environments. Software agents, or simply agents, are usually defined as autonomous software entities, with various degrees of intelligence, capable of exhibiting both reactive and pro-active behaviour in order to satisfy their design goals. From the point of e-learning and tutoring systems harvester and pedagogical agents are of the special research interest. Harvester agents are in charge of collecting learning material from online, often heterogeneous repositories and success depends on the quality and standards of teaching material representation. The main goals of pedagogical agents are to motivate and guide students through the learning process, by asking questions and proposing solutions. This chapter presents a possible trend in use of intelligent agents for personalised learning within tutoring system. Some possibilities of the use of several kinds of agents in a stand-alone e-learning architecture are proposed.

A recent trend in the field of e-learning and tutoring systems is to utilize agent technology and develop and use different kinds of agents. Despite the fact that it is possible to design, develop and use different kinds of sophisticated agents (Ivanović et al. 2014) two categories of software agents are of the special interest in this domain: harvester and pedagogical agents.

Software agents, or simply agents, are usually defined as autonomous software entities, with various degrees of intelligence, capable of exhibiting both reactive and pro-active behaviour in order to satisfy their design goals (Bădică et al. 2011). From the point of e-learning and tutoring systems harvester and pedagogical agents are of the special research interest. Harvester agents are in charge of collecting learning material from online, often heterogeneous repositories and success depends on the quality and standard of teaching material representation. The core properties of the agent technology, such as parallel and distributed execution, mobility, and inter-agent communication, can bring significant benefits to the harvesting process, in terms of speed and efficiency (Badica and Badica 2009; De la Prieta and Gil 2010).

(Heidig and Clarebout 2011) define pedagogical agents as "lifelike characters presented on a computer screen that guide users through multimedia learning

© Springer International Publishing Switzerland 2017   43
A. Klašnja-Milićević et al., *E-Learning Systems*,
Intelligent Systems Reference Library 112, DOI 10.1007/978-3-319-41163-7_5

environments". The main goals of these agents are to motivate and guide students through the learning process, by asking questions and proposing solutions (Heller and Procter 2010).

In following sections, possibilities of using several kinds of agents in a stand-alone e-learning architecture will be presented.

5.1   Some Existing Agent Based Systems

Recently, several interesting approaches that use software agents in e-learning systems are reported. Intelligent agents are incorporated in ABITS (Capuano et al. 2000), MathTuthor (Frigo et al. 2005), and Educ-MAS (Gago et al. 2009), in order to improve the course recommendation results.

Harvester agents are proposed in Sharma and Gupta (2010) to utilize the parallel and distributed execution of the agent in order to optimize the web crawling process. Another system that relies on harvester agents is presented in Baran et al. (2007)—Agent Based Search System (ABSS). It is intended to improve the quality of search query results not only harvesting heterogeneous remote learning object repositories, but also tracking changes in them. Harvested learning objects are stored in local repositories to be searched by a sophisticated agent-based search module. Similarly (Barcelos et al. 2011) have developed an agent-based federated catalog of learning objects (AgCAT)—an efficient system for searching and retrieving learning objects.

Both of these systems use sophisticated harvesting agents to retrieve the best-suited learning objects, i.e. they are sophisticated search engines and enable their users to pull the data using search queries.

On the other hand other kinds of agents are studied in Heidig and Clarebout (2011). They have provided an interesting analysis of 39 studies related to the effects of pedagogical agents onto the learning outcome. The initial conclusion is that only 5 studies have detected positive effects of using pedagogical agents, but majority of them did not use a control group that learn without the support of agents. Therefore, the suggestion is for the researchers to better examine additional elements of learning like the environment, domains, levels of design, etc. that lead to the beneficial use of pedagogical agents.

SmartEgg is a web-based pedagogical agent that assists students in learning SQL (Mitrovic et al. 2007). It is integrated into an intelligent e-learning system named SQL-Tutor. The agent includes a visual representation with animated gestures, and can express different behaviours. Initial evaluation has shown that the usage of SmartEgg agent has a significant positive impact onto the student's motivation (Mitrovic et al. 2004).

Although many existing systems incorporate either type of agents, there have been no attempts to efficiently integrate both harvesting and motivational-level agents. HAPA (HArvester and Pedagogical Agent-based e-learning system) is the concept of helpful and misleading pedagogical agents in the same environment and with the common goal (Ivanović et al. 2015). HAPA is designed as general architecture that incorporates agents to facilitate higher quality e-learning activities. But, first prototype implemented to prove the concept is in the area of programming and it helps learners in studying essential elements of the programming language Java. HAPA consists of three main components:

- harvester agents,
- classifier module and
- pair of pedagogical agents.

The task of harvester agents is a collection of the appropriate learning material from the web. Their results are further delivered to the Classifier module. This module performs automatic classification of individual learning objects that will be proposed to a learner. A pair of specially de-signed pedagogical agents (helpful and misleading) interacts with students and helps them comprehend the learning material.

5.2   HAPA System Overview

HAPA system is currently standalone e-learning system which helps students in learning and especially in solving programming problems by writing appropriate programs. At a later stage *HAPA* could be included as a component in other tutoring systems devoted to learning programming languages. The goal of the system is to be incorporate existing e-learning systems.

A high-level overview of the system architecture is outlined in Fig. 5.1. *HAPA* includes several important components: harvester agents, the *Classifier* module,

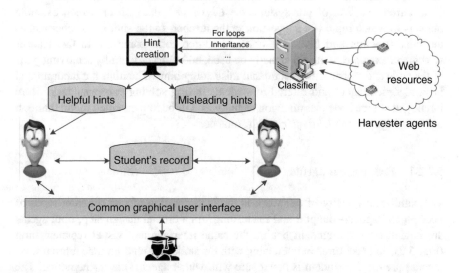

Fig. 5.1 A high-level overview of the *HAPA* system

repositories of helpful and misleading hints, and pedagogical agents. The functioning of each component and their mutual interactions are described in more details in the following sub-sections (Ivanović et al. 2015).

5.2.1 Harvesting and Classifying the Learning Material

HAPA is currently mainly focused on the *code completion* type of tasks, in which students are expected to fill-in missing parts of the program. The student is given a code snippet, and then requested to complete the source code according to task specification. This kind of tasks is well-suited for testing but also for improving the student's programming skills.

Initial activity of constructing the code completion tasks is collection of additional learning material. This material is collected by the harvester agents. For the purpose of *HAPA*, the learning material consists of *Java* source code examples. With the abundance of these examples available on the web, harvester agents have been implemented as web crawlers.

After processing the current page, the agent continues the harvesting process on all pages linked from the current one, and so on up to a predefined depth. Many agents can be deployed on a computer cluster and perform the harvesting in parallel and maintained in order to avoid duplicate work. The harvested learning material is fed into the *Classifier* module, which automatically associates each *Java* source code example with appropriate lecture topic i.e. each example is assigned to the appropriate lecture topic, such as "for loops", "inheritance", "input/output", and so on. On the other hand, the teacher is able to analyse and focus on examples of a particular interest, i.e. those that are directly attached to the lecture topic in question.

In current version of our system, the *Classifier*'s decision on which example belongs to which topic is a suggestion to the teacher. In the end, the teacher makes the final selection and filters the obtained source code examples. In fact teacher selects the examples that will actually be used, and process them by removing parts of the code and constructing useful and misleading hints. The hints are incorporated in pedagogical agent and offered to students during solving particular task. More intelligent source code classification techniques will be implemented in the future in order to improve the *Classifier*'s performance.

5.2.1.1 Pedagogical agents

The significant novelty of our work is the incorporation of two different types of pedagogical agents—helpful and misleading. As a crucial design step, both agents are hidden from the student behind the same interface and visual representation (Fig. 5.2), and take turns in interacting with the student at random time intervals. As a consequence, the student is never sure with which agent (s)he is interacting. The main idea of this approach is rather simple: to motivate students not to trust the

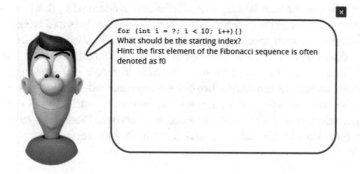

Fig. 5.2 Visual representation of a Pedagogical *HAPA* agent

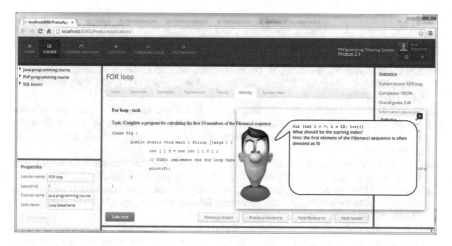

Fig. 5.3 Learning session supported by visual Pedagogical agent

agent's hints blindly and stimulate them to critically analyse the problem in question and the proposed hint, and independently decide on the proper solution.

Recent trend in educational systems is to represent pedagogical agents as life-like, animated characters so we also decided to implement simple visual Pedagogical *HAPA* agent and let students decide if they will activate it or not (Fig. 5.3). In some situations visual agents can distract the user/student from concentrating on the problem in question, and in the extreme case, it may negatively affect his/her willingness to use the system.

Both pedagogical agents are capable of adapting to each individual student. Agents track a set of information about the student, including his/her personal data, the ratio of correct and incorrect solutions to each code completion problem, and the student's grade for each lecture topic. Based on these data agents can intervene if the student's success rate becomes unsatisfactory. Additionally, it will repeat the appropriate code completion tasks until a certain success threshold is reached.

We believe that such approach where students do not know if agent gives correct or wrong directions and hints is challenging. It can additionally motivate students to critically think and assess their knowledge. Also in future real working environments students will face different helpful but also malicious colleagues who will maybe suggest them wrong and unacceptable procedures and steps. So we would like to put students in unexpected situations and motivate them to be cautious and reassess their knowledge and skills and not blindly believe in external hints and suggestions.

References

Badica, A., & Badica, C. (2009). Specification and verification of an agent-based auction service. *Information Systems Development: Towards a Service Provision Society* (pp. 239–248). doi:10.1007/b137171_25.

Bădică, C., Budimac, Z., Burkhard, H. D., & Ivanović, M. (2011). Software agents: Languages, tools, platforms. *Computer Science and Information Systems, 8*(2), 255–296. doi:10.2298/CSIS110214013B.

Baran, R., Zeja, A., Orzechowski, T., Dziech, A., & Lutwin, M. (2007). The data collection module of the agent based search system (ABSS). In: *Proceedings of Webist 2007–3rd International Conference on Web Information Systems and Technologies*, (Vol. IT, pp. 451–454).

Barcelos, C. F., Gluz, J. C., & Vicari, R. M. (2011). An Agent-based Federated Learning Object Search Service. *Interdisciplinary Journal of E-Learning and Learning Objects, 7*.

Capuano, N., Marsella, M., & Salerno, S. (2000). ABITS: An agent based intelligent tutoring system for distance learning. In: *The Electronic Library* Vol. 20, (pp. 134–142). http://doi.org/10.1108/02640470210424473.

De la Prieta, F., & Gil, A. B. (2010). A multi-agent system that searches for learning objects in heterogeneous repositories. In: *Trends in Practical Applications of Agents and Multiagent Systems*. (pp. 355–362).

Frigo, L., Cardoso, J., & Bittencourt, G. (2005). Adaptive interaction in intelligent tutoring systems. *Proceedings of CIAH-2005, International Workshop on Combining Intelligent and Adaptive Hypermedia Methods/Techniques in Web-based Education Systems in Salzburg-Autriche* (pp. 33–38).Salzburg-Autriche Austria.

Gago, I. S. B., Werneck, V. M. B., & Costa, R. M. (2009). Modeling an educational multi-agent system in maSE. In *Active Media Technology* (pp. 335–346). Springer.

Heidig, S., & Clarebout, G. (2011). Do pedagogical agents make a difference to student motivation and learning? *Educational Research Review*. http://doi.org/10.1016/j.edurev.2010.07.004.

Heller, B., & Procter, M. (2010). Animated pedagogical agents & immersive worlds: Two worlds colliding. Emerging Technologies in Distance Education.

Ivanović, M., Mitrović, D., Budimac, Z., Jerinić, L., & Bădică, C. (2015). HAPA: Harvester and pedagogical agents in e-learning environments. *International Journal of Computers Communications & Control, 10*(2), 200–210.

Ivanović, M., Mitrović, D., Budimac, Z., Vesin, B., & Jerinić, L. (2014). Different roles of agents in personalized programming. *New Horizons in Web Based Learning, 7697*, 161–170.

Mitrovic, A., Martin, B., & Suraweera, P. (2007). Intelligent tutors for all: The constraint-based approach. *IEEE Intelligent Systems*, *22*(4), 38–45. http://doi.org/10.1109/MIS.2007.74.

Mitrovic, A., Suraweera, P., Martin, B., & Weerasinghe, A. (2004). DB-suite: Experiences with Three Intelligent, Web-based Database Tutors. *Journal of Interactive Learning Research, 15* (4), 409–432.

Sharma, S., & Gupta, J. P. (2010). A novel architecture of agent based crawling for OAI resources. *International Journal of Computer Science and Engineering, 2*(4), 1190–1195.

Chapter 6
Recommender Systems in E-Learning Environments

Abstract Recommender system can be defined as a platform for providing recommendations to users based on their personal likes and dislikes. These systems use a specific type of information filtering technique that attempt to recommend information items (movies, music, books, news, Web pages, learning objects, and so on.) to the user. Recommender systems strongly depend on the context or domain they operate in, and it is often not possible to take a recommendation strategy from one context and transfer it to another context or domain. Personalized recommendation can help learners to overcome the information overload problem, by recommending learning resources according to learners' habits and level of knowledge. The first challenge for designing a recommender component for e-learning systems is to define the learners and the purpose of the specific context or domain in a proper way. This chapter provides an overview of techniques for recommender systems, folksonomy and tag-based recommendation to assist the reader in understanding the material which follows in subsequent chapters.

Recomender systems (RS) strongly depend on the context or domain they operate in, and it is often not possible to take a recommendation strategy (Drachsler et al. 2009) from one context and transfer it to another context or domain. The first challenge for designing a RS is to define the learners and the purpose of the specific context or domain in a proper way (McNee et al. 2006). Learning process includes three components: learners, teachers/instructors, and learning materials. From a teacher's point of view, teaching is an activity to deliver information and skill to learners with some goals to be achieved. From the learners' point of view, learning is an activity to acquire information from teacher to achieve goals set by the teacher. Learners, with their prior knowledge, acquire new information from the teacher. Here, social constructivism paradigm can help learners learn collaborative and sharing knowledge with each other. Basically, knowledge which is needed to be achieved according to the course, mainly does not influence how many Learning

© Springer International Publishing Switzerland 2017
A. Klašnja-Milićević et al., *E-Learning Systems*,
Intelligent Systems Reference Library 112, DOI 10.1007/978-3-319-41163-7_6

Objects (LOs) the learners have read, but how relevant are LOs that have retrieved and learned. A learner who has high prior knowledge according to the course is different from other learners who have low prior knowledge.

In a virtual classroom, teachers provide resources such as text, multimedia and simulations, and moderate and animate discussions. Remote learners are encouraged to peruse the resources and participate in activities. However, it is very difficult and time consuming for educators to thoroughly track and assess all the activities performed by all learners on all tools. Moreover, it is hard to evaluate the structure of the course content and its effectiveness on the learning process. Resource providers do their best to structure the content assuming its efficacy (Zaïane and Luo 2001). When instructors put together an on-line course, they may compile interactive course notes, simulations, demos, exercises, quizzes, asynchronous forums, chat tools, Web resources, etc. This amalgam of on-line hyperlinked material could form a complex structure that is difficult to navigate. Hence, personalization features are needed, which adaptively facilitate learner in monitoring their learning progress and provide any resources or learning material, that's suitable to what they need.

6.1 Recommendations and Recommender Systems

The information on the Web is increasing far more quickly than people can cope with. Learners are forced to review a number of choices before they discover what they need. This is often time consuming and frustrating. Given today's fast paced lifestyle, a slow and careful search for the elusive item of choice is surely not a sustainable option. People would rather look at items that are customized to their interests and preferences. Personalized recommendation (Resnick and Varian 1997) can help people to overcome the information overload problem, by recommending items according to users' interests.

Recommender systems can be defined as a platform for providing recommendations to users based on their personal likes and dislikes. These systems use a specific type of information filtering (IF) technique that attempt to recommend information items (movies, music, books, news, Web pages, learning objects, etc.) to the user (Ricci et al. 2011). To handle this the user's profile is compared to some reference characteristics. These characteristics may be from the information item (the content-based approach) or the user's social environment (the collaborative filtering approach) (Adomavicius and Tuzhilin 2005a).

Typically, RSs apply personalization techniques, considering that different users have different preferences and different information needs (Herlocker et al. 2004). In order to generate personalized recommendations that are tailored to the user's specific needs, recommender systems must collect personal preference information, e.g., the user's history of purchase, click-stream data, demographic information, and so forth. Traditionally, expressions of preference of users for products are generally called ratings. Two different types of ratings are distinguished.

1. *Explicit ratings.* Users are required to explicitly specify their preference for any particular item, usually by indicating their extent of appreciation on 5 or 7-point Likert scales (Ziegler 2013).
2. *Implicit ratings.* Explicit ratings require additional efforts on users. Consequently, users often tend to avoid the burden of explicitly stating their preferences and either leave the system or rely upon "free-riding" (Avery and Zeckhauser 1997). Alternatively, gathering preference information from observations of user behaviour is less intrusive (David 1997).

A ratings database consists of pairs of users and items rated by them, along with additional information such as timestamps, etc. Given such a ratings database and a set of preliminary ratings by a new user, the basic requirement of a recommendation system is to recommend the largest and the most significant set of recommendations to the new user (Miller 2003). The largest set of recommendations refers to the maximum number of items in the item database that could be recommended to the user. The significant items are the ones that the new user will be more likely to rate in the future.

Based on the nature of reference characteristics, two broad categories of information filtering for computing recommendations have emerged: content-based filtering, and collaborative filtering (Goldberg et al. 1992).

Content-based recommender systems recommend items "by comparing representations of content contained in an item to representations of content that interests the user" (Malone et al. 1987). In most cases, a keyword profile is created. Apparently, this works well in text domains, but not in domains where there is not much content associated with the items or where a computer cannot easily analyse this content. Relying on rich descriptions, content-based recommender systems need significant knowledge engineering efforts to create substantial metadata for the items. Content-based systems "form profiles for each user independently" (Basilico and Hofmann 2004). Even if two items were in two neighbour categories that people normally like both of them, the item from category A would never be recommended to the user if (s)he only rated items from category B. This problem is often addressed by introducing some unpredictability. Also, the user has to rate a suitable number of items before a content-based recommender system can really comprehend the user's preferences and present the user with trustworthy recommendations. Therefore, a new user, having very few ratings, would not be able to get accurate recommendations.

Collaborative filtering systems compute profile similarity between the target user and the other users "by comparing users' opinions of items" (Balabanović and Shoham 1997). Profile similarity is usually computed by comparing rating-vectors with various distance metrics, e.g. Pearson correlation or cosine similarity. They supply the user with the items (s)he will most likely be interested in, either one single item or a "ranked list of items"—usually referred as *top-N-item* (Cosley et al. 2002; McLaughlin and Herlocker 2004). In contrast to content-based systems, recommender systems based on collaborative filtering can provide the user with unexpected but fitting recommendations that do not have anything in common with afore rated items. Collaborative filtering is a very successful methodology in almost every domain—especially "where multi-value ratings are available" (McLaughlin and Herlocker 2004). However, they suffer from two key problems: sparsity and first-rater problem. As most users only rate a small portion of all items, it is highly difficult to find users with "significantly similar ratings." Furthermore an item cannot be recommended before one user has rated it. This can be the case if the item has newly been introduced to the system (Melville et al. 2002).

A number recommendation systems use a *hybrid approach* by combining collaborative and content-based methods, which helps to avoid certain limitations of content and collaborative-based systems (Balabanović and Shoham 1997; Basu et al. 1998; Claypool et al. 1999; Pennock and Horvitz 1999). Different ways to combine collaborative and content-based methods into a hybrid recommender system can be classified as follows (Adomavicius and Tuzhilin 2005a):

1. implementing collaborative and content-based methods separately and combining their predictions,
2. incorporating some content-based characteristics into a collaborative approach,
3. incorporating some collaborative characteristics into a content-based approach, and
4. constructing a general unifying model that incorporates both content-based and collaborative characteristics.

According to (Breese et al. 1998), algorithms for collaborative recommendations can be grouped into two general classes: memory-based (or heuristic-based) and model-based.

Memory-based algorithms (Breese et al. 1998; Resnick et al. 1994; Shardanand and Maes 1995) essentially are heuristics that utilize the entire database of user preferences when computing recommendations. These algorithms tend to be simple to implement and require little training cost. They can also easily take new preference data into account. However, their online performance tends to be slow as the size of the user and item sets grow, which makes these algorithms as stated in the literature unsuitable in large systems. One workaround is to only consider a subset of the preference data in the calculation, but doing this can reduce both recommendation quality and the number of items that can be recommended due to data being omitted from the calculation. Another solution is to perform as much of the computation as possible in an offline setting. However, this may make it difficult to add new users to the system on a real-time basis, which is a basic necessity of most online systems. Furthermore, the storage requirements for the pre-computed data could be high.

Model-based algorithms (Billsus and Pazzani 1998; Goldberg et al. 2001; Hofmann 2003) use the collection of ratings to learn a model, which is then used to make rating predictions. Often, the model building process is time-consuming and is only used periodically. The model is compact and can generate recommendations very quickly. The disadvantage of model-based algorithms is adding new users, items, or preferences, which can be the same as re-computing the entire model.

The most important difference between collaborative model-based techniques and heuristic-based approaches is that the model-based techniques calculate utility predictions based not on some ad hoc heuristic rules, but, rather, based on a model learned from the underlying data using statistical and machine learning techniques (Adomavicius and Tuzhilin 2005a). A method combining memory-based and model-based approaches was proposed in Pennock et al. (2000). It was empirically confirmed that the use of this approach can afford better recommendations than pure memory-based and model-based collaborative approaches.

Over the past several years there has been much research done on recommendation technologies which use a variety of statistical, machine learning, information retrieval, and other techniques that have significantly advanced early recommender systems, collaborative and content-based heuristics. As was discussed above, recommender systems can be classified as (1) content-based, collaborative, or hybrid (based on the recommendation approach used), and (2) heuristic-based or model-based (based on the types of recommendation techniques used for the rating estimation). These two orthogonal dimensions are used to classify the recommender systems research in the 2×3 matrix, presented in Table 6.1 (Adomavicius and Tuzhilin 2005a).

The recommendation techniques explained in this chapter have performed well in several applications, including the ones for recommending books, CDs, news articles or movies (Marlin 2003; Rosset et al. 2002) and some of these methods are used in the "industrial-strength" recommender systems, such as the ones developed at Amazon,[1] MovieLens,[2] and Last.fm.[3] However, both collaborative and content-based methods have certain limitations. Recommender systems can be extended in several ways that include improving the understanding of users and items, incorporating the contextual information into the recommendation process, sustaining multicriteria ratings, and providing more flexible and less disturbing types of recommendations (Adomavicius and Tuzhilin 2005a).

---

[1]http://www.amazon.com.

[2]http://www.movielens.umn.edu.

[3]http://www.last.fm.

Table 6.1 Classification of RS research (Adomavicius and Tuzhilin 2005b)

Recommendation Approach	Recommendation technique	
	Heuristic based	Model based
Content - based	Commonly used techniques:	Commonly used techniques:
	TF-IDF (information retrieval)	Bayesian classifiers
	Clustering	Clustering
	Representative research examples:	Decision trees
	Lang 1995	Representative research examples:
	Balabanović, Shoham 1997	Pazzani and Billsus 1997
	Pazzani and Billsus 1997	Mooney et al. 1998
		Mooney and Roy 1999
		Billsus and Pazzani 1998, 1999
		Zhang et al. 2002
Collaborative	Commonly used techniques:	Commonly used techniques:
	Nearest neighbour	Bayesian networks
	(cosine, correlation)	Clustering
	Clustering	Artificial neural networks
	Graph theory	Linear regression
	Representative research examples:	Probabilistic models
	Resnick et al. 1994	Representative research examples:
	Hill et al. 1994	Billsus and Pazzani 1998
	Shardannand and Maes 1995	Breese et al. 1998
	Breese et al. 1998	Goldberg et al. 2001
	Nakamura and Abe 1998	Ungar and Foster 1998
	Aggarwal et al. 1999	Chien and George 1999
	Delgado and Ishii 1999	Getoor and Sahami 1999
	Pennock and Horwitz 1999	Pennock and Horwitz 1999
	Sarwar et al. 2001	Pavlov and Pennock 2002
		Shani et al. 2002
		Hofmman 2003, 2004
Hybrid	Combining content—based and collaborative components by:	Combining content—based and collaborative components by:
	Linear combination of predicted ratings	Incorporating one component as a part of the model for the other
	Various voting schemes	Building one unifying model
	Incorporating one component as a part of the heuristic for the other	Representative research examples:
		Soboroff and Nicholas 1999
	Representative research examples:	Basu et al. 1998
	Balabanović and Shoham 1997	Condiff et al. 1999
	Pazzani 1999	Popescul et al. 2007
	Billsus and Pazzani 1998	Schein et al. 2002
	Claypool et al. 1999	Ansari et al. 2000
	Good et al. 1999	
	Train and Cohen 2000	

6.2   The Most Important Requirements and Challenges for Designing a Recommender System in E-Learning Environments

A RS in e-learning environments utilizes information about learners and learning activities (LA) and recommend items such as papers, Web pages, courses, lessons and other learning resources which meet the pedagogical characteristics and interests of learners (Drachsler et al. 2008). Such a RS could provide recommendations to online learning materials or shortcuts. Those recommendations are based on previous learners' activities or on the learning styles of the learners that are discovered from their navigation patterns. To design an effective RS in e-learning environments, it is important to understand specific learners' characteristics (Drachsler et al. 2008; García et al. 2009):

1. Learner's goal or learner's task is a feature related with the context of a learner's activities in educational system rather than with the learner as an individual. Depending on the kind of system, it can be the goal of the activity (in application systems), a search goal (in information retrieval systems), and a problem solving or learning goal (in educational (e-learning) systems). In all of these cases the goal is an answer to the question "Why is the learner using the system and what does the learner actually want to achieve?" Learner's goal is the most unpredictable learner feature: almost always it changes from session to session and often can change several times within one session. In some systems it is reasonable to distinguish local or low-level goals which can change quite often and general or high level goals and tasks which are more stable. For example, in educational systems the knowledge acquisition is a high-level goal, while the problem-solving goal is a low-level goal which changes from one educational problem to another several times within a session.

2. Prior learner's knowledge of the subject represents one of the most important feature of the learner for adaptive educational systems. Almost all adaptive presentation techniques rely on the learner's knowledge as a source of adaptation. Related to the first point, we also need to know if the learners already have any prior knowledge about what they want to learn. The proficiency level of the learning activity should fit the proficiency level of the learner (prior knowledge). The learners may want to reach the learning goals on specific competence levels, like beginner, advanced or expert levels. Learner's knowledge is a variable for a particular learner. This means that an adaptive educational system which relies on learner's knowledge has to recognize the changes in the learner's knowledge state and update the learner model accordingly.

3. Background and experience are two features of the learner which are similar to user's knowledge of the subject but functionally differ from it. Learner's background can be presented as all the information related to the learner's previous experience outside the subject of the educational system, which is relevant enough to be considered. This includes the learner's profession,

experience of activities in related areas, as well as the learner's point of view
and perspective. According to learner's experience it could be identified how
familiar is the learner with the structure of the similar learning environments and
how easy can the learner navigate in it. This is not the same as learner's
knowledge of the subject (Vassileva 1998). Sometimes, the learner who is
generally quite familiar with the subject itself is not familiar at all with the
structure of educational system. Vice versa, the learner can be quite familiar
with the structure of the educational system without deep knowledge of the
subject. One more reason to differentiate experience from knowledge level is the
existence of an adaptive navigation technique (Pérez et al. 1995; Vassileva
1998) which relies on this feature of the learner.

4. Learner preferences. For different reasons the learner can prefer some links more
   than others and some parts of a page more than others. These preferences can be
   absolute or relative, i.e., dependent from the current link, goal and current
   context in general. Learner's preferences differ from other learner model components in several aspects. Unlike other components, the preferences cannot be
   deduced by the system. The learner has to inform the system directly or indi-
   rectly (by a simple feedback) about such preferences. It looks more close to
   adaptability then to adaptivity.

5. Learner group models accumulate preferences of a specific group of learners
   (such as a research laboratory). A group model is a nice starting model for a new
   member of the group. Group models are important also for collaborative
   activities. It is very hard to collaborate when collaborators use individual learner
   models and thus have different adapted views on the same subject.

6. Rated learning activities (LAs). The aggregated ratings of the learning activities
   as awarded by other learners can provide valuable information (the rated
   learning activities). Learners with the same learning goal or similar study time
   per week could benefit from the ratings received from more advanced learners.
   Nearly all potential learning activities are unknown to the learners. Learners are
   (by definition) not able to rate learning activities in advance, because if they
   already knew them, they would no longer be potential learning activities.
   Moreover, the learners will at least have to read through a learning activity
   before they are able to rate it. Many people are able to rate movies because they
   have heard or read about it, or have already seen the movie. In the domain of
   learning, however, it is unlikely that a learner will already be familiar with
   certain learning activities. Consequently, it is less of a problem for 'movie
   lovers' to rate movies in advance to specify a profile than it is for learners to rate
   learning activities in advance. Requiring learners to rate an initial set of learning
   activities, as in movielens.org, does not, therefore, seem feasible. Other mechanisms to specify a learner profile have to be devised. Even for the learners with
   the same interests, we may need to recommend different learning activities,
   depending on the individual proficiency levels, learning goals and context. For
   instance, the learners with no prior knowledge in a specific domain should be
   advised to study basic learning activities first, while more advanced learners
   should be advised to continue with more specific learning activities.

7. Learning paths. Beginning learners could benefit from historical information about the successful study behaviour of the more advanced learners in the same learning network, in the same learning paths. From the learning activities which are frequently positively rated and their sequence, the most popular learning paths will emerge. The most successful learning paths with regard to efficiency and effectiveness could be recommended.

8. Learning strategies. RS in e-learning would benefit if we apply the learning strategies derived from educational psychology research (Koper and Olivier 2004). Such strategies could use pedagogical rules as guiding principles for recommendation, like 'go from simple to more complex tasks' or 'gradually decrease the amount of contact and direct guidance'. This entails taking into account the metadata about specific learning activities, but not the actual design of the specific learning activities themselves.

E-learning systems should be able to recognize and exploit these learners' characteristics serve as guidelines for framework design and platform implementation for a good RS for e-learning (Angehrn et al. 2001; Savidis et al. 2007; Zaïane and Luo 2001).

- *A good RS should be highly personalized.* Relevant learning materials should be chosen and presented to learners or researchers based on learner's learning style, interests, preferences, current activities, etc.
- *A good RS should recommend materials at the appropriate time and location.* A good RS should deliver relevant learning materials to learner at the most appropriate time and locations to facilitate learners' acquisition of knowledge and skills.
- *A good RS should support non-disruptive view of experience.* Non-disruptive means that learners have the option to either follow or discount relevant materials based on their learning needs.
- *A good RS should be socially situated.* A good RS should be able to recognize and exploit the learners' social networks, role models, levels of trust and influence, etc. RS should also help the learners to recognize their knowledge acquisition process in the context of the group.
- *A good RS should include the adoption phase.* A good RS should be able to monitor, understand and model the different phases of adoption of the knowledge by the learner. In particular it includes the phases in which the new concepts are experimented with, evaluated, internalized and finally applied.
- *A good RS should support the continuous learning process.* A good RS should support just-in-time learning, by better analysing their current and future activities. Also it should provide motivational support and stimulation.
- *A good RS should provide high level of interactivity.* A good RS should provide very active, cognitive and diverse mode of interaction with the learner in the form of a rich choice of interaction strategies.
- *A good RS should provide appropriate course materials according to learners' learning style.* Each person learns differently and needs to develop his/her own

learning skills in his/her own way. Learners have different backgrounds, strengths and weaknesses, interests, ambitions, senses of responsibility, levels of motivation, and approaches to studying and learning. For example, different learners prefer different presentation forms: some prefer multimedia contents (simulations, presentations, graphical material and hypertext documents); while others prefer traditional Web pages (questionnaires, exercises, research studies).

6.3 Recommendation Techniques for RS in E-Learning Environments—A Survey of the State-of-the-Art

Personalized recommendation approaches are first proposed in e-commerce area for product purchase (Balabanović and Shoham 1997; Resnick and Varian 1997), which help consumers to find products they would like to purchase by creating a list of recommended products for each given consumer (Cheung et al 2003; Schafer et al. 2001). Literature review shows that there are also many researchers who have attempted to adopt recommender systems to e-learning environments. For example, (Shen and Shen 2005) described a mechanism focused on how to organize the learning materials based on domain ontology which can guide the learning resources recommendation according to learning status. A multi-attribute assessment method is proposed in Lu (2004) to justify a learner's need and deployed a fuzzy matching method to find suitable learning contents to best perform each learner need. Research paper (Luo et al. 2002) presented a method to organize components and courseware using the hierarchy and association rules of the concepts, which can recommend the relative contents to learners and also can help them to control the learning schedule. However, most of these methods missing one important issue in e-learning RS, that is, the natural learning behaviour is not lonely but interactive which relying on friends, classmates, lecturers, and other sources to make the choices for learning.

Designers and instructors, when devising the on-line structure of the course and course material, have a navigation pattern in mind and assume all on-line learners would follow a consistent path; the path put out in the design and materialized by some hyperlinks. Learners, however could follow different paths generating a variety of sequences of learning activities. Often some sequences are not the optimum sequences, and probably not the sequence intended by the designer. Instructors are in desperate need for non-intrusive and automatic ways to get objective feedback from learners in order to better follow the learning process and appraise the on-line course structure effectiveness. On the learner's side, it would be very useful if the system could automatically guide the learner's activities and intelligently recommend on-line activities or resources that would favour and improve the learning. The automatic recommendation could be based on the teacher's intended sequence of navigation in the course material, or, more interestingly, based on navigation patterns of other successful learners. For example, during the learning process, a learner read a useful material, summarized what (s)he has learned or got the answer of a typical question, some learners with similar learning status will likely need these resources.

E-learning system uses different recommendation techniques in order to suggest online learning activities to learners, based on their preferences, knowledge and the browsing history of other learners with similar characteristics. RSs assist the natural process of relying on friends, classmates, lecturers, and other sources for making the choices of learning (Lu 2004). In the educational setting, these recommendation systems can be classified according to their field of application or focus (Romero et al. 2007):

1. learner-centered (Gaudioso et al 2003; Zaíane 2002), in order to suggest good learning experiences for the learners in accordance to their preferences, needs and level of knowledge; and
2. teacher-centered, with the aim of helping the teachers and/or authors of the e-learning systems to improve the functionalities or performances of these systems based on learner information (W Chen and Wasson 2003; Romero et al. 2003). Some other examples of educational applications of these systems are: obtaining more feedback about teaching; finding out more about how learners learn on the Web; evaluating learners in terms of their browsing patterns; classifying learners into groups; or restructuring the contents of the website in order to personalize the course.

Each recommendation strategy has its own strengths and weaknesses. According to set of the most important requirements for a good RS in e-learning environment, have been explored and defined in the previous section, in the remainder of this section we present a survey of the state-of-the-art in RSs for e-learning systems. We identify challenges and various limitations for each traditional recommendation method, then consider some tag-based profiling approaches for extending their capabilities.

6.3.1   Collaborative Filtering Approach

Collaborative systems track past actions of a group of learners to make a recommendation for individual members of the group (Tan et al. 2008). Based on the assumption that learners with similar past behaviours (rating, browsing, or learning path) have similar interests, a collaborative filtering system recommends learning objects the neighbours of the given learner have liked.

This approach relies on a history record of all learner interests such as can be inferred from their ratings of the items (learning objects/learning actions) on a website. Rating can be explicit (explicit ratings or customer satisfaction questionnaires) or implicit (from the studying patterns or click-stream behaviour of the learners). The proportion of actual studying hours to the total hours of the course is recorded as the implicit rating scores, and transformed to corresponding explicit rating scores, from 1 to 5. The learners' rating scores can be given in a m * n

Table 6.2 Learner's rating matrix

	O_1	...	O_k	...	O_n
I_1	$R_{1,\,1}$	...	$R_{1,\,k}$	...	$R_{1,\,n}$
...					
I_1	$R_{j,\,1}$	...	$R_{j,\,k}$	...	$R_{j,\,n}$
...					
I_1	$R_{m,\,1}$	...	$R_{m,\,k}$	...	$R_{m,\,n}$

matrix, as it is shown in Table 6.2, where $L = \{I_1, I_2, \ldots, I_m\}$ is a list of m learners, $O = \{o_1, o_2, \ldots, o_n\}$ is the list of n learning objects, and $R_{j,k}$ gives the rating of object O_k, given by learner j. Also, it can be rating of object o_k given by intelligent tutoring system for learner j. There exists a distinguished learner $I_a \in L$ called the active learner for whom the task of collaborative filtering algorithm is to find learning object likeliness.

The neighbourhood formation scheme usually uses Pearson correlation or cosine similarity as a measure of proximity (Resnick et al. 1994; Shardanand and Maes 1995).

An exploratory study of a recommender system, using collaborative filtering to support (virtual) learners in a learning network, has been reported in Koper (2005). The author simulated rules for increasing/decreasing motivation and some other disturbance factors in learning networks, using the Netlogo tool. Closely related to this study is an experiment reported in Janssen et al. (2007). The authors offered to learners a similar recommendation system. The recommendations did not take personal characteristics of learners (or possible 'matching errors') into account. Another system implemented by Soonthornphisaj et al. (2006) allows all learners to collaborate their expertise in order to predict the most suitable learning materials to each learner. This smart e-learning system applies the collaborative filtering approach that has an ability to predict the most suitable documents to the learner. All learners have the chance to introduce new material by uploading the documents to the server or pointing out the Web link from the Internet and rate the currently available materials.

One of the first attempts to develop a collaborative filtering system for learning resources has been the Altered Vista (AV) system (Recker et al. 2003; Recker and Walker 2003; Walker et al. 2004). The AV system (Walker et al. 2004) uses a database in which a learner evaluations of learning resources are stored. Learners can browse the reviews of others and can get personalized learning resource recommendations from the system. AV does not aim to support learners directly by giving them feedback on their work. Instead, AV provides an indirect learning support in which suitable learning tools are recommended. The team working on AV explored several relevant issues, such as the development of non-authoritative metadata to store learner-provided evaluations (Recker and Walker 2003), the design of the system and the review scheme it uses (Walker et al. 2004), as well as results from pilot and empirical studies from using the system to recommend to the members of a community both interesting resources and people with similar tastes and beliefs. A survey-based evaluation of AV showed a predominant positive feedback, but also identified issues with the system's incentive and with regard to privacy (Walker et al. 2004).

Another system of the educational collaborative filtering applications is the Web-based PeerGrader (PG) (Gehringer 2001; Lynch et al. 2006). The purpose of this tool is to help learners improve their skills by reviewing and evaluating solutions of their fellow learners blindly. PG works in the following way:

- the learners get a task list and each learner chooses a task,
- the learners submit their solutions to the system, where they are read by another learner who then provides feedback in form of textual comments,
- the authors modify their solutions based on the comments they have received, and re-submit their modified solutions again to the system, where they will be reviewed by other learners, then the solutions' authors grade each review with respect to whether it was helpful or not.
- finally, the system calculates grades for all learner solutions.

One of PG's strengths is to provide learners with high-quality feedback also in ill-defined homework tasks that do not have clear-cut gold standard solutions (such as design problems). This kind of feedback could not be generated automatically. A disadvantage is the time required for the system to work effectively: due to the complexity of the reviewing process and the textual comments, the evaluation of a single learner answer is very time consuming. This may cause learner drop-outs and deadline problems (Lynch et al. 2006). Also, studies with PG revealed problems with getting feedback of high quality. An evaluation of subjective usefulness showed that the system was appreciated by its users (Lynch et al. 2006), yet a systematic comparison of PG scores to expert grades has not been conducted.

A newer Web-based collaborative filtering system, the Scaffolded Writing and Rewriting in the Discipline (SWoRD) system (Cho et al. 2006; Cho and Schunn 2007) addresses the problem of writing homework in the form of a long text, which cannot be reviewed in detail by a teacher for time reasons. Because of this, learners do often not receive any detailed feedback on their solutions at all. Having such feedback, it would be beneficial for learners, since they could use it to improve their future work. To address this problem, SWoRD relies on peer reviews and implements an algorithm that follows the typical journal publication and reviewing process. An evaluation showed that the participants benefitted from multi-peers' feedback more than from single-peer's or single expert's feedback (Cho and Schunn 2007).

A different approach is used by the LARGO system (Pinkwart et al. 2006), where learners create graphs of US Supreme Court oral arguments. Within LARGO, collaborative scoring is employed to assess the quality of a "decision rule" that a learner has included in his diagram. Since this assessment involves interpretation of legal argument in textual form, it cannot be automated reasonably. While the overall LARGO system has been tested in law schools and shown to help lower-aptitude learners (Pinkwart et al. 2007), empirical studies to test the educational effectiveness of the specific collaborative scoring components have not been conducted.

Rule-Applying Collaborative Filtering (RACOFI) Composer system (Anderson et al. 2003; Lemire et al. 2005; Lemire 2005) combines two recommendation approaches by integrating a collaborative filtering engine, that works with ratings that learners provide for learning resources, with an inference rule engine that is mining association rules between the learning resources and using them for recommendation. RACOFI studies have not yet assessed the pedagogical value of the recommender, nor do they report some evaluation of the system by learners.

Manouselis and (Manouselis and Costopoulou 2007) tried a typical, neighborhood-based set of collaborative filtering algorithms in order to support learning object recommendation. The examined algorithms have been multi-attribute ones, allowing the recommendation service to consider multi-dimensional ratings that learners provide on learning resources. The performance of the same algorithms is changing, depending on the context where testing takes place. The results from the comparative study of the same algorithms in an e-commerce and a e-learning setting (Manouselis et al. 2011) have led to the selection of different algorithms from the same set of candidate ones.

In summary, the relatively few educational systems with collaborative filtering components have an underlying algorithm to determine solution quality based on collaborative scoring. Yet, existing systems are often specialized for a particular application area such as legal argumentation (LARGO), writing skills training (SWoRD), or educational resource recommendation (AV), or they involve a rather complicated and long-term review process (SWoRD, PG).

The collaborative filtering (CF) based techniques, in general, suffer from several limitations. Two serious limitations with quality evaluation are: the sparsity problem and the "cold-start" problem (Lu 2004). The sparsity problem occurs when available data is insufficient for identifying similar learners or items (neighbours) due to an immense amount of learners and items (Sarwar et al. 2001). It is difficult for collaborative filtering based recommender systems to precisely compute the neighbourhood and identify the learning objects to be recommended even though learners are very active, each individual has only expressed a rating on a very small portion of the items (Linden et al. 2003). Also, a severe problem is the cold start problem (first-rater), which occurs when a new learner/learner object is introduced and thus has no previous ratings information available (Massa and Avesani 2004). With this situation, the system is generally unable to make high quality recommendations.

The CF-based techniques rely heavily on explicit learner input (e.g., previous customers' rating/ranking of products), which is either unavailable or considered intrusive. With sparsity of such learner input, the recommendation precision and quality drop significantly. This is because without good and trusted ratings entered by the learners, recommendations become useless and untrustworthy. To recommend learning activities or learning objects it is better to use real past activities (history logs) by learners as input for their profiles. Also, in the case of intelligent tutoring system, collaborative filtering approach can be carried out according to ratings (grades) for learners' knowledge level, provided by the tutoring system.

6.3.2 Content-Based Techniques

Content-based techniques recommend items (learning objects/learning actions) similar to the ones the learners preferred in the past. They base their recommendations on individual information and ignore contributions from other learners (Billsus and Pazzani 1998). In content-based systems, items are described by a common set of attributes. Learner's preferences are predicted by considering the association between the item ratings and the corresponding item attributes. Therefore, learner can receive proper recommendations without help from other learners. Content-based techniques can be classified into two different categories (Aguzzoli et al 2002; Schmitt and Bergmann 1999; Wilson et al. 2003):

1. Case based reasoning (CBR) techniques and
2. Attribute—based techniques.

Case based reasoning techniques recommend items with the highest correlation to items the learner liked before. Case-based reasoning is useful to keep the learner informed about aimed learning goals. These techniques are domain-independent, do not require content analysis and the quality of the recommendation improves over time when the learners have rated more items. The disadvantage of the new learner problem also states to case-based reasoning techniques. Nevertheless, specific disadvantages of case-based reasoning are overspecialization and sparsity, because only items that are highly correlated with the learner profile or interest can be recommended. Through case-based reasoning the learner is limited to a set of items that are similar to the items (s)he already knows (Adomavicius and Tuzhilin 2005a).

Recent research papers present different facets of CBR in teaching or learning process. Pixed (Project Integrating eXperience in Distance Learning), which is an adaptive hypermedia ontology-based system implements case based reasoning method (Heraud et al. 2004). The Pixed approach assumes positions of a learner as a kind of expert of her/his own learning skills, or at least as a real practitioner of his own practices. The learner builds her/his knowledge by interacting with the learning environment, trying to benefit as much as possible from the available educational activities. Learning is considered as a problem-solving task. The goal is to learn a specific concept proposed in the domain knowledge ontology. The way to reach this goal is one particular path among the different available educational activities linked to that ontology. (Sørmo and Aamodt 2002) propose building "a cognitive model of how humans solve problems in the domain and use this model in attempting to solve the problem, both from the point of view of the current learner (using the learner model) and of an expert (represented by an expert model)". The case-based reasoner has to evaluate the learner's solution and to explain why s/he does or does not fit the observed features of the problem. (Funk and Conlan 2003) make research more closely related to Pixed. Their goal is the same: to use learner feedback in order to adapt the learning environment. The learner feedback can be exploited in two ways: direct feedback exploitation during the learning process, in the form of learners' comments, and feedback exploitation by authors and tutors after the learning process in order to integrate it into the proposed courses, by comparing the learners' result with the result of other cases. The authors associate CBR with filtering techniques by attempting to create learner profiles taking into account different feedbacks. (Elorriaga and Fernández-Castro 2000) propose to use CBR to deploy an instructional planner which adapts the sequences observed in logs in order to create instructional sequences for a complete course. In Heraud et al. (2004), a case—based reasoning system was developed to offer navigational guidance to the learner. It is based on past user's interaction logs and it includes a model describing learning sessions.

Attribute–based techniques recommend items based on the matching of their attributes to the learner profile. Attributes could be weighted for their importance to learner. Adding new LAa or learners to the network will not cause any problem. Attribute-based techniques are sensitive to changes in the profiles of the learners (Drachsler et al. 2008). They can always control the personalized RS by changing their profile or the relative weight of the attributes. A description of needs in their profile is mapped directly to available LA. A serious disadvantage is that an attribute-based recommendation is static and not able to learn from the network behaviour. That is the reason why highly personalized recommendation cannot be achieved. Attribute-based techniques work only with information that can be described in categories. Media types, like audio and video, first need to be classified to the topics in the profile of the learner. This requires category modelling and maintenance which could raise serious limitations for learning environments. Also the overspecialization can be a problem, especially if learners do not change their profile. Attribute-based recommendations are useful to handle the 'cold-start' problem because no behaviour data about the learners is needed. Attribute-based techniques can directly map characteristics of learners (like learning goal, prior knowledge, and available study time) to characteristics of LA (Drachsler et al. 2007). There are several applications that tackle attribute—based techniques problems such as prediction and visualization. Attribute–based Ant Colony System (AACS) (Yang and Wu 2009) uses a method of finding learning objects that would be suitable for a learner based on the most frequent learning trails followed by the previous learners. The system updates the trails pheromones from different knowledge levels and different styles of learners to create a powerful and dynamic learning object search mechanism. There are three prerequisites for achieving this:

1. the adaptive learning portal knows the learner's attributes which include the learner's knowledge level and learning style
2. the learner's attributes and learning object's attributes which have been annotated by teacher or content providers
3. matching the relationships between learners and learning object.

6.3.3 Association Rule Mining

Association rule mining techniques (Agrawal and Srikant 1995) are one of the most popular ways of representing discovered knowledge and describe a close correlation between frequent items in a database. An association rule consists of an antecedent (left-hand side) and a consequent (right-hand side). The intersection between the antecedent and the consequent is empty. An:

$$X \Rightarrow Y$$

type association rule expresses a close correlation between items (attribute-value) in a database (Zheng et al. 2001). Most association rule mining algorithms require the user to set at least two thresholds, one of minimum support and the other of minimum confidence. The support S of a rule is defined as the probability that an entry satisfies both X and Y. Confidence is defined as the probability an entry has satisfies Y when it satisfies X. Therefore the aim is to find all the association rules that satisfy certain minimum support and confidence restrictions, with parameters specified by the user. Therefore, the user must have a certain amount of expertise in order to find the right support and confidence settings to achieve the best rules.

Association rule mining has been applied to e-learning systems in order to intelligently recommend on-line learning activities to learners, based on the actions of previous learners which can improve course content navigation as well as to assist the on-line learning process (Arenas-García et al. 2007).

Count the learners' browsing records, learning path and testing grades and finding out the connection between learning objects, association rule can be used to calculate the learning profiles of the new learners and perform the following tasks:

- building recommender agents for on-line learning activities or shortcuts (Zaíane 2002),
- automatically leading the learner's activities and intelligently recommend on-line learning activities or shortcuts in the course Web site to the learners (Lu 2004),
- identifying attributes of performance inconsistency between various groups of learners (Minaei-Bidgoli et al. 2004),
- discovering interesting learner's usage information in order to provide feedback to course author (Romero et al. 2004),
- finding out the relation among the learning materials from a large amount of educational material data (Lu et al. 2003),
- finding learners' mistakes that are often occur together (Merceron and Yacef 2004),
- optimizing the content of an e-learning portal by determining the content of most interest to the learner (Ramli 2005),
- deriving useful patterns to help educators and instructors evaluating and interpreting on-line course activities (Zaiane 2002), and

- personalizing e-learning based on comprehensive usage profiles and a domain ontology (Markellou et al. 2005).

Most of the subjective approaches involve learner participation in order to express, in accordance to his or her previous knowledge, which rules are of interest. Hence, subjective measures are becoming increasingly important (Silberschatz and Tuzhilin 1996). Some suggested subjective measures (Liu et al. 2000) are:

- *Unexpectedness*: Rules are interesting if they are unknown to the learner or contradict the learner's knowledge.
- *Actionability*: Rules are interesting if learners can do something with them.

There are several specific research papers about the application of association rule mining and recommender systems in e-learning systems. Association rules for classification applied to e-learning (Castro et al. 2007), have been investigated in the areas of learning recommendation systems (Chu et al. 2003; Zaïane 2002), learning material organization (Tsai et al. 2001), learner learning assessments (Hwang et al. 2003; Kumar 2005; Okamoto and Matsui 2003; Silva and Vieira 2001), course adaptation to the learners' behaviour (Hsu et al. 2003; Markellou et al. 2005), and evaluation of educational Web sites (Machado and Becker 2003).

(Wang et al. 2002) developed a portfolio analysis tool based on associative material clusters and sequences among them. This knowledge allows teachers to study the dynamic browsing structure and to identify interesting or unexpected learning patterns. (Minaei-Bidgoli et al. 2004) propose mining interesting contrast rules for Web-based education systems. Contrast rules help one to identify attributes characterizing patterns of performance difference between various groups of learners. (Markellou et al. 2005) propose an ontology-based framework and discover association rules, using the Apriori algorithm. The role of the ontology is to determine which learning materials are more suitable to be recommended to the learner. (Jia Li and Zaïane 2004) use recommender agents for recommending online learning activities or shortcuts in a course Web site based on a learner's access history. (Romero et al. 2004) propose to use grammar-based genetic programming with multi-objective optimization techniques for discovering useful association rules from learner's usage information. (Merceron and Yacef 2004) use association rules and symbolic data analysis, as well as traditional SQL queries to mining learner data captured from a Web-based tutoring tool. Their goal is to find mistakes that often occur together. (Freyberger et al. 2004) use association rules to determine what operation to perform on the transfer model that predicts a learner's success.

Apriori algorithm (Agrawal et al. 1993) is a prominent algorithm for mining frequent itemsets for Boolean association rules. In Apriori algorithm, it is time-consuming that the database has been scanned for many times. Therefore, many algorithms, like the DIC algorithm (Brin et al. 1997), DHP algorithm (Park et al. 1995) and AprioriTid algorithm (Agrawal et al. 1993), etc., are proposed successively to improve the performance.

Association rule mining and frequent pattern mining were applied in (Zaïane 2002) to extract useful patterns that might help teacher, educational managers, and

Web masters to evaluate and understand on-line course activities. A similar approach can be found in Minaei-Bidgoli et al. (2004), where distinguish rules, defined as sets of conjunctive rules describing patterns of performance difference between groups of learners, were used. A computer-assisted approach to diagnosing learner learning problems in science courses and offer learners advice was presented in Hwang et al. (2003), based on the concept effect relationship (CER) model, a specification of the association rules technique.

A hypermedia learning environment with a tutorial component was described in (Costabile De Marsico et al. 2005). It is called Logiocando and targets children of the fourth level of primary school (9–10 years old). It includes a tutor module, based on if-then rules, that emulates the teacher by providing suggestions on how and what to study. In Okamoto and Matsui (2003) it can be found the description of a learning process assessment method that resorts to association rules, and the well-known ID3 DT learning method. A framework that employ Web usage mining to support the validation of learning site designs was defined in Machado and Becker (2003), applying association and sequence techniques (Srivastava et al. 2000).

In Markellou et al. (2005), a framework for personalized e-learning based on aggregate usage profiles and domain ontology were presented, and a combination of Semantic Web and Web mining methods was used. The Apriori algorithm for association rules was applied to capture relations among URL references based on the navigational patterns of learners. A test result feedback (TRF) model that analyses the relationships between learner learning time and the corresponding test results was introduced in Hsu et al. (2003). The objective was twofold: on the one hand, developing a tool for supporting the tutor in reorganizing the course material; on the other, a personalization of the course tailored to the individual learner needs. The approach was based on association rules mining. A rule-based mechanism for the adaptive generation of problems in Intelligent Tutoring System (ITS) in the context of Web-based programming tutors was proposed in Kumar (2005). In Hwang et al. (2003), a Web-based course recommendation system, used to provide learners with suggestions when having trouble in choosing courses, was described. The approach integrates the Apriori algorithm with graph theory.

Some of the main drawbacks of association rule algorithms are (García et al. 2007):

- association rule mining algorithms normally discover a huge quantity of rules and do not guarantee that all the rules found are relevant,
- the used algorithms have too many parameters for somebody non expert in data mining and
- the obtained rules are far too many, most of them non-interesting and with low comprehensibility.

In order to provide better recommendations, and to be able to use recommender systems in more complex types of e-learning environments, most of the methods reviewed in this subsection would need significant extensions. Therefore, we consider some tag-based profiling approaches for extending their capabilities.

References

Adomavicius, G., & Tuzhilin, A. (2005a). Personalization technologies. *Communications of the ACM.* http://doi.org/10.1145/1089107.1089109.

Adomavicius, G., & Tuzhilin, A. (2005b). Toward the next generation of recommender systems: A survey of the state-of-the-art and possible extensions. *IEEE Transactions on Knowledge and Data Engineering, 17*(6), 734–749.

Agrawal, R., Imielinski, T., & Swami, A. (1993). Database mining: A performance perspective. *Knowledge and Data Engineering, IEEE Transactions on, 5*(6), 914–925.

Agrawal, R., & Srikant, R. (1995). Mining sequential patterns. In *Data Engineering, 1995. Proceedings of the Eleventh International Conference on* (pp. 3–14).

Aguzzoli, S., Avesani, P., & Massa, P. (2002). Collaborative case-based recommender systems. In *Advances in Case-Based Reasoning* (pp. 460–474). Springer.

Anderson, M., Ball, M., Boley, H., Greene, S., Howse, N., McGrath, S., & Lemire, D. (2003). Racofi: A rule-applying collaborative filtering system.

Angehrn, A., Nabeth, T., Razmerita, L., & Roda, C. (2001). K-InCA: using artificial agents to help people learn and adopt new behaviours. In *Advanced Learning Technologies, 2001. Proceedings. IEEE International Conference on* (pp. 225–226).

Arenas-García, J., Meng, A., Petersen, K. B., Lehn-Schioler, T., Hansen, L. K., & Larsen, J. (2007). Unveiling music structure via plsa similarity fusion. In *Machine Learning for Signal Processing, 2007 IEEE Workshop on* (pp. 419–424).

Balabanović, M., & Shoham, Y. (1997). Fab: Content-based, collaborative recommendation. *Communications of the ACM.* http://doi.org/10.1145/245108.245124.

Basilico, J., & Hofmann, T. (2004). Unifying collaborative and content-based filtering. In *Proceedings of the twenty-first international conference on Machine learning* (p. 9).

Basu, C., Hirsh, H., Cohen, W., & others. (1998). Recommendation as classification: Using social and content-based information in recommendation. In *AAAI/IAAI* (pp. 714–720).

Billsus, D., & Pazzani, M. J. (1998). Learning Collaborative Information Filters. In *ICML* (Vol. 98, pp. 46–54).

Breese, J. S., Heckerman, D., & Kadie, C. (1998). Empirical analysis of predictive algorithms for collaborative filtering. In *Proceedings of the Fourteenth conference on Uncertainty in artificial intelligence* (pp. 43–52).

Brin, S., Motwani, R., Ullman, J. D., & Tsur, S. (1997). Dynamic itemset counting and implication rules for market basket data. In *ACM SIGMOD Record* (Vol. 26, pp. 255–264).

Castro, F., Vellido, A., Nebot, À., & Mugica, F. (2007). Applying data mining techniques to e-learning problems. In *Evolution of teaching and learning paradigms in intelligent environment* (pp. 183–221). Springer.

Chen, W., & Wasson, B. (2003). Coordinating collaborative knowledge building. *International Journal of Computers and Applications, 25*(1), 1–10.

Cheung, K. W., Kwok, J. T., Law, M. H., & Tsui, K. C. (2003). Mining customer product ratings for personalized marketing. *Decision Support Systems, 35*(2), 231–243. http://doi.org/10.1016/S0167-9236(02)00108-2.

Cho, K., & Schunn, C. D. (2007). Scaffolded writing and rewriting in the discipline: A web-based reciprocal peer review system. *Computers & Education, 48*(3), 409–426.

Cho, K., Schunn, C. D., & Wilson, R. W. (2006). Validity and reliability of scaffolded peer assessment of writing from instructor and student perspectives. *Journal of Educational Psychology, 98*(4), 891.

Chu, K.-K., Chang, M., & Hsia, Y.-T. (2003). Designing a course recommendation system on web based on the students' course selection records. In *World conference on educational multimedia, hypermedia and telecommunications* (Vol. 2003, pp. 14–21).

Claypool, M., Gokhale, A., Miranda, T., Murnikov, P., Netes, D., & Sartin, M. (1999). Combining content-based and collaborative filters in an online newspaper. In *Proceedings of ACM SIGIR workshop on recommender systems* (Vol. 60).

Cosley, D., Lawrence, S., & Pennock, D. M. (2002). REFEREE: An open framework for practical testing of recommender systems using ResearchIndex. In *Proceedings of the 28th international conference on Very Large Data Bases* (pp. 35–46).

Costabile, M. F., De Marsico, M., Lanzilotti, R., Plantamura, V. L., & Roselli, T. (2005). On the usability evaluation of e-learning applications. In *System Sciences, 2005. HICSS'05. Proceedings of the 38th Annual Hawaii International Conference on* (p. 6b–6b).

David, N. (1997). Implicit Rating and Filtering. In *5th DELOS Workshop on Filtering and Collaborative Filtering (ERCIM), Budapest, Hungary* (pp. 31–36).

Drachsler, H., Hummel, H., & Koper, R. (2007). Recommendations for learners are different: Applying memory-based recommender system techniques to lifelong learning.

Drachsler, H., Hummel, H. G. K., & Koper, R. (2008). Personal recommender systems for learners in lifelong learning networks: the requirements, techniques and model. *International Journal of Learning Technology, 3*(4), 404–423.

Drachsler, H., Hummel, H. G. K., & Koper, R. (2009). Identifying the goal, user model and conditions of recommender systems for formal and informal learning. *Journal of Digital Information, 10*(2).

Elorriaga, J. A., & Fernández-Castro, I. (2000). Using case-based reasoning in instructional planning. towards a hybrid self-improving instructional planner. *International Journal of Artificial Intelligence in Education (IJAIED), 11*, 416–449.

Freyberger, J., Heffernan, N., & Ruiz, C. (2004). Using association rules to guide a search for best fitting transfer models of student learning. In *Workshop on analyzing student-tutor interactions logs to improve educational outcomes at ITS conference* (pp. 1–10).

Funk, P., & Conlan, O. (2003). Using case-based reasoning to support authors of adaptive hypermedia systems. In *AH2003: workshop on adaptive hypermedia and adaptive web-based systems* (pp. 113–120).

García, P., Amandi, A., Schiaffino, S., & Campo, M. (2007). Evaluating Bayesian networks' precision for detecting students' learning styles. *Computers & Education, 49*(3), 794–808.

García, E., Romero, C., Ventura, S., & De Castro, C. (2009). An architecture for making recommendations to courseware authors using association rule mining and collaborative filtering. *User Modeling and User-Adapted Interaction, 19*(1–2), 99–132.

Gaudioso, E., Santos, O. C., Rodríguez, A., & Boticario, J. G. (2003). A proposal for modeling a collaborative task in a web-based collaborative learning environment. In *International Conference on User Modeling* (pp. 70–79).

Gehringer, E. F. (2001). Electronic peer review and peer grading in computer-science courses. *ACM SIGCSE Bulletin, 33*(1), 139–143.

Goldberg, D., Nichols, D., Oki, B. M., & Terry, D. (1992). Using collaborative filtering to weave an information tapestry. *Communications of the ACM, 35*(12), 61–70.

Goldberg, K., Roeder, T., Gupta, D., & Perkins, C. (2001). Eigentaste: A constant time collaborative filtering algorithm. *Information Retrieval, 4*(2), 133–151.

Heraud, J.-M., France, L., & Mille, A. (2004). Pixed: An ITS that guides students with the help of learners' interaction log. In *International conference on intelligent tutoring systems, workshop analyzing student tutor interaction logs to improve educational outcomes. Maceio, Brazil* (pp. 57–64).

Herlocker, J. L., Konstan, J. A., Terveen, L. G., & Riedl, J. T. (2004). Evaluating collaborative filtering recommender systems. *ACM Transactions on Information Systems.* http://doi.org/10.1145/963770.963772.

Hofmann, T. (2003). Collaborative filtering via gaussian probabilistic latent semantic analysis. In *Proceedings of the 26th annual international ACM SIGIR conference on Research and development in informaion retrieval* (pp. 259–266).

Hsu, H.-H., Chen, C.-J., & Tai, W.-P. (2003). Towards error-free and personalized Web-based courses. In *Advanced Information Networking and Applications, 2003. AINA 2003. 17th International Conference on* (pp. 99–104).

Hwang, G.-J., Hsiao, C.-L., & Tseng, J. C. R. (2003). A computer-assisted approach to diagnosing student learning problems in science courses. *J. Inf. Sci. Eng., 19*(2), 229–248.

Janssen, J., den Berg, B., Tattersall, C., Hummel, H., & Koper, R. (2007). Navigational support in lifelong learning: enhancing effectiveness through indirect social navigation. *Interactive Learning Environments, 15*(2), 127–136.

Koper, R. (2005). Increasing learner retention in a simulated learning network using indirect social interaction. *Journal of Artificial Societies and Social Simulation, 8*(2).

Koper, R., & Olivier, B. (2004). Representing the learning design of units of learning. *Educational Technology and Society.*

Kumar, A. (2005). Rule-based adaptive problem generation in programming tutors and its evaluation. In *The 12th international conference on artificial intelligence in education. July* (pp. 18–22).

Lemire, D. (2005). Scale and translation invariant collaborative filtering systems. *Information Retrieval, 8*(1), 129–150.

Lemire, D., Boley, H., McGrath, S., & Ball, M. (2005). Collaborative filtering and inference rules for context-aware learning object recommendation. *Interactive Technology and Smart Education, 2*(3), 179–188.

Li, J., & Zaïane, O. R. (2004). Combining usage, content, and structure data to improve web site recommendation. In *E-Commerce and Web Technologies* (pp. 305–315). Springer.

Linden, G., Smith, B., & York, J. (2003). Amazon. com recommendations: Item-to-item collaborative filtering. *Internet Computing, IEEE, 7*(1), 76–80.

Liu, B., Hsu, W., Chen, S., & Ma, Y. (2000). Analyzing the subjective interestingness of association rules. *Intelligent Systems and Their Applications, IEEE, 15*(5), 47–55.

Lu, J. (2004). Personalized e-learning material recommender system. In *International conference on information technology for application* (pp. 374–379).

Lu, J., Yu, C. S., & Liu, C. (2003). Learning style, learning patterns, and learning performance in a WebCT-based MIS course. *Information and Management, 40*(6), 497–507. http://doi.org/10.1016/S0378-7206(02)00064-2.

Luo, S., Sha, S., Shen, D., & Jia, W. (2002). Conceptual network based courseware navigation and web presentation mechanisms. In *Advances in Web-Based Learning* (pp. 81–91). Springer.

Lynch, C., Ashley, K., Aleven, V., & Pinkwart, N. (2006). Defining ill-defined domains; a literature survey. In *Proceedings of the workshop on intelligent tutoring systems for ill-defined domains at the 8th international conference on intelligent tutoring systems* (pp. 1–10).

Malone, T. W., Grant, K. R., Turbak, F. A., Brobst, S. A., & Cohen, M. D. (1987). Intelligent information-sharing systems. *Communications of the ACM, 30*(5), 390–402.

Machado, L. dos S., & Becker, K. (2003). Distance education: A web usage mining case study for the evaluation of learning sites. In *Advanced Learning Technologies, 2003. Proceedings. The 3rd IEEE International Conference on* (pp. 360–361).

Manouselis, N., & Costopoulou, C. (2007). Analysis and classification of multi-criteria recommender systems. *World Wide Web, 10*(4), 415–441. http://doi.org/10.1007/s11280-007-0019-8.

Manouselis, N., Drachsler, H., Vuorikari, R., Hummel, H., & Koper, R. (2011). Recommender Systems in Technology Enhanced Learning. In *Recommender Systems Handbook* (pp. 387–415). http://doi.org/10.1007/978-0-387-85820-3.

Markellou, P., Mousourouli, I., Spiros, S., & Tsakalidis, A. (2005). Using semantic web mining technologies for personalized e-learning experiences. *Proceedings of the Web-Based Education, 461–826.*

Marlin, B. M. (2003). Modeling User Rating Profiles For Collaborative Filtering. In *NIPS* (pp. 627–634).

Massa, P., & Avesani, P. (2004). Trust-aware collaborative filtering for recommender systems. In *On the Move to Meaningful Internet Systems 2004: CoopIS, DOA, and ODBASE* (pp. 492–508). Springer.

McLaughlin, M. R., & Herlocker, J. L. (2004). A collaborative filtering algorithm and evaluation metric that accurately model the user experience. In *Proceedings of the 27th annual international ACM SIGIR conference on Research and development in information retrieval* (pp. 329–336).

McNee, S. M., Riedl, J., & Konstan, J. A. (2006). Making recommendations better: An analytic model for human-recommender interaction. In *CHI'06 extended abstracts on Human factors in computing systems* (pp. 1103–1108).

Melville, P., Mooney, R. J., & Nagarajan, R. (2002). Content-boosted collaborative filtering for improved recommendations. In *AAAI/IAAI* (pp. 187–192).

Merceron, A., & Yacef, K. (2004). Mining student data captured from a web-based tutoring tool: Initial exploration and results. *Journal of Interactive Learning Research, 15*(4), 319.

Miller, B. (2003). *Toward a personalized recommender system.* PhD thesis, University of Minnesota–Twin Cities.

Minaei-Bidgoli, B., Tan, P.-N., & Punch, W. (2004). Mining interesting contrast rules for a web-based educational system. In *Machine Learning and Applications, 2004. Proceedings. 2004 International Conference on* (pp. 320–327).

Okamoto, T., & Matsui, T. (2003). Knowledge Discovery from Learning History Data and and its Effective Use for Learning Process Assessment under the e-learning Environment. In *Society for Information Technology & Teacher Education International Conference* (Vol. 2003, pp. 3141–3144).

Park, J. S., Chen, M.-S., & Yu, P. S. (1995). *An effective hash-based algorithm for mining association rules* (Vol. 24). ACM.

Pennock, D. M., & Horvitz, E. (1999). Analysis of the axiomatic foundations of collaborative filtering. *Ann Arbor, 1001*, 42110–48109.

Pennock, D. M., Horvitz, E., Lawrence, S., & Giles, C. L. (2000). Collaborative filtering by personality diagnosis: A hybrid memory-and model-based approach. In *Proceedings of the Sixteenth conference on Uncertainty in artificial intelligence* (pp. 473–480).

Pérez, T., Lopistéguy, P., Gutiérrez, J., & Usandizaga, I. (1995). HyperTutor: From hypermedia to intelligent adaptive hypermedia. In *ED-MEDIA'95, World conference on educational multimedia and hypermedia* (pp. 529–534).

Pinkwart, N., Aleven, V., Ashley, K., & Lynch, C. (2006). Toward legal argument instruction with graph grammars and collaborative filtering techniques. In *Intelligent Tutoring Systems* (pp. 227–236).

Pinkwart, N., Aleven, V., Ashley, K., & Lynch, C. (2007). Evaluating legal argument instruction with graphical representations using largo. *Frontiers In Artificial Intelligence And Applications, 158*, 101.

Ramli, A. A. (2005). Web usage mining using apriori algorithm: UUM learning care portal case. In *International conference on knowledge management, Malaysia* (pp. 1–19).

Recker, M. M., & Walker, A. (2003). Supporting" word-of-mouth" social networks through collaborative information filtering. *Journal of Interactive Learning Research, 14*(1), 79.

Recker, M. M., Walker, A., & Lawless, K. (2003). What do you recommend? Implementation and analyses of collaborative information filtering of web resources for education. *Instructional Science, 31*(4–5), 299–316.

Resnick, P., Iacovou, N., Suchak, M., Bergstrom, P., & Riedl, J. (1994). GroupLens: An open architecture for collaborative filtering of netnews. In *Proceedings of the 1994 ACM conference on Computer supported cooperative work* (pp. 175–186).

Resnick, P., & Varian, H. R. (1997). Recommender systems. *Communications of the ACM.* http://doi.org/10.1145/245108.245121.

Ricci, F., Rokach, L., & Shapira, B. (2011). *Introduction to Recommender Systems Handbook. Recommender Systems Handbook.* http://doi.org/10.1007/978-0-387-85820-3_1.

Romero, C., Ventura, S., & De Bra, P. (2004). Knowledge discovery with genetic programming for providing feedback to courseware authors. *User Modelling and User-Adapted Interaction, 14*(5), 425–464. http://doi.org/10.1007/s11257-004-7961-2.

Romero, C., Ventura, S., De Bra, P., & De Castro, C. (2003). Discovering prediction rules in AHA! courses. In *User Modeling 2003* (pp. 25–34). Springer.

Romero, C., Ventura, S., Delgado, J. A., & De Bra, P. (2007). Personalized links recommendation based on data mining in adaptive educational hypermedia systems. In *Creating New Learning Experiences on a Global Scale* (pp. 292–306). Springer.

Rosset, S., Neumann, E., Eick, U., Vatnik, N., & Idan, Y. (2002). Customer lifetime value modeling and its use for customer retention planning. In *Proceedings of the eighth ACM SIGKDD international conference on Knowledge discovery and data mining* (pp. 332–340).

Sarwar, B., Karypis, G., Konstan, J., & Riedl, J. (2001). Item-based collaborative filtering recommendation algorithms. *Proceedings of the 10th ...*, *1*, 285–295. http://doi.org/10.1145/371920.372071.

Savidis, A., Grammenos, D., & Stephanidis, C. (2007). Developing inclusive e-learning and e-entertainment to effectively accommodate learning difficulties. *Universal Access in the Information Society, 5*(4), 401–419.

Schafer, J. Ben, Konstan, J., & Riedl, J. (2001). E-commerce recommendation applications. *Applications of Data Mining to Electronic ...*, 115–153. http://doi.org/10.1007/978-1-4615-1627-9_6.

Schmitt, S., & Bergmann, R. (1999). Applying case-based reasoning technology for product selection and customization in electronic commerce environments. In *12th Bled Electronic Commerce Conference* (Vol. 273).

Shardanand, U., & Maes, P. (1995). Social Information Filtering: Algorithms for Automating "Word of Mouth." In *ACM Conference on Human Factors in Computing Systems (CHI)* (Vol. 1, pp. 210–217). http://doi.org/10.1145/223904.223931.

Shen, L., & Shen, R. (2005). Ontology-based learning content recommendation. *International Journal of Continuing Engineering Education and Life Long Learning, 15*(3–6), 308–317.

Silberschatz, A., & Tuzhilin, A. (1996). What makes patterns interesting in knowledge discovery systems. *Knowledge and Data Engineering, IEEE Transactions on, 8*(6), 970–974.

Silva, D. R., & Vieira, M. T. P. (2001). An Ongoing assessment model in distance learning.

Soonthornphisaj, N., Rojsattarat, E., & Yim-Ngam, S. (2006). Smart e-learning using recommender system. In *Computational intelligence* (pp. 518–523). Springer.

Sørmo, F., & Aamodt, A. (2002). Knowledge communication and CBR. In *ECCBR Workshops* (pp. 47–60).

Srivastava, J., Cooley, R., Deshpande, M., & Tan, P.-N. (2000). Web usage mining: Discovery and applications of usage patterns from web data. *ACM SIGKDD Explorations Newsletter, 1*(2), 12–23.

Tan, H., Guo, J., & Li, Y. (2008). E-learning recommendation system. In *Computer Science and Software Engineering, 2008 International Conference on* (Vol. 5, pp. 430–433).

Tsai, C.-J., Tseng, S.-S., & Lin, C.-Y. (2001). A two-phase fuzzy mining and learning algorithm for adaptive learning environment. In *Computational Science-ICCS 2001* (pp. 429–438). Springer.

Vassileva, J. (1998). DCG + GTE: Dynamic Courseware Generation with Teaching Expertise. *Instructional Science, 26*(3–4), 317–332.

Walker, A., Recker, M. M., Lawless, K., & Wiley, D. (2004). Collaborative information filtering: A review and an educational application. *International Journal of Artificial Intelligence in Education, 14*(1), 3–28.

Wang, D., Bao, Y., Yu, G., & Wang, G. (2002). Using page classification and association rule mining for personalized recommendation in distance learning. In *Advances in Web-Based Learning* (pp. 363–374). Springer.

Wilson, D. C., Smyth, B., & Sullivan, D. O. (2003). Sparsity reduction in collaborative recommendation: A case-based approach. *International Journal of Pattern Recognition and Artificial Intelligence, 17*(05), 863–884.

Yang, Y. J., & Wu, C. (2009). An attribute-based ant colony system for adaptive learning object recommendation. *Expert Systems with Applications, 36*(2), 3034–3047.

Zaiane, O. (2002). Building a Recommender Agent for e learning Systems Computers in Education. *Citeulike.org*.

Zaíane, O. R. (2002). Building a recommender agent for e-learning systems. In *Computers in Education, 2002. Proceedings. International Conference on* (pp. 55–59).

Zaïane, O. R., & Luo, J. (2001). Towards evaluating learners' behaviour in a web-based distance learning environment. In *Advanced Learning Technologies, 2001. Proceedings. IEEE International Conference on* (pp. 357–360).

Zheng, Z., Kohavi, R., & Mason, L. (2001). Real world performance of association rule algorithms. In *Proceedings of the seventh ACM SIGKDD international conference on Knowledge discovery and data mining* (pp. 401–406).

Ziegler, C.-N. (2013). *Social Web Artifacts for Boosting Recommenders*. Springer.

Chapter 7
Folksonomy and Tag-Based Recommender Systems in E-Learning Environments

Abstract Collaborative tagging is technique, highly employed in different domains, which is used for automatic analysis of users' preferences and recommendations. To improve recommendation quality, metadata such as content information of items has typically been used as additional knowledge. With the increasing reputation of the collaborative tagging systems, tags could be interesting and provide useful information to enhance algorithms for recommender systems. Besides helping user to organize his/her personal collections, a tag also can be regarded as a user's personal opinion expression, while tagging can be considered as implicit rating or voting on the tagged information resources or items. The overview, presented in this chapter includes descriptions of content-based recommender systems, collaborative filtering systems, hybrid approach, memory-based and model-based algorithms, features of collaborative tagging that are generally attributed to their success and popularity, as well as a model for tagging activities and tag-based recommender systems.

Collaborative tagging is a technique, highly employed in different domains, which is used for automatic analysis of users' preferences and recommendations. To improve recommendation quality, metadata such as content information of items has typically been used as additional knowledge. With the increasing reputation of the collaborative tagging systems, tags could be interesting and useful information to enhance algorithms for recommender systems. Collaborative tagging systems allow users to upload their resources, and to label them with arbitrary words, so-called tags (Golder and Huberman 2006). The systems can be distinguished according to what kind of resources they supported. Flickr,[1] for instance, allows the sharing of photos, del.icio.us[2] the sharing of bookmarks, CiteULike[3] and Connotea[4] the sharing of bibliographic references, and 43Things[5] even the sharing of goals in

---

[1]http://www.flickr.com.
[2]http://www.del.icio.us.
[3]http://www.citeulike.org.
[4]http://www.connotea.org.
[5]http://www.43things.com.

© Springer International Publishing Switzerland 2017
A. Klašnja-Milićević et al., *E-Learning Systems*,
Intelligent Systems Reference Library 112, DOI 10.1007/978-3-319-41163-7_7

private life. Essentially, all these systems are very similar. Once a user is logged in, (s)he can add a resource to the system, and assign arbitrary tags. The collection of all his assignments is his personomy, the collection of all personomies constitutes the folksonomy. The user can explore his personomy, as well as the personomies of the other users, in all dimensions: for a given user one can see all resources (s)he had uploaded, together with the tags (s)he had assigned to them (Jäschke et al. 2007). Besides helping user to organize his/her personal collections, a tag also can be regarded as a user's personal opinion expression, while tagging can be considered as implicit rating or voting on the tagged information resources or items (Liang et al. 2008). Thus, the tagging information can be used to make recommendations. The rest of this chapter will review in more detail the collaborative tagging systems, folksonomy and tag-based RSs, relevant to the results of the research presented in this monography. Section 7.1 provides comprehensive survey of the state-of-the-art in collaborative tagging systems and folksonomy. Section 7.2 presents a model for tagging activities. Tag-based recommender systems and approaches for extension and collecting tags are described in Sect. 7.3.

7.1   Comprehensive Survey of the State-of-the-Art in Collaborative Tagging Systems and Folksonomy

Collaborative tagging is the practice of allowing users to freely attach keywords or tags to content (Golder and Huberman 2006). Collaborative tagging is most useful when there is nobody in the "librarian" role or there is simply too much content for a single authority to classify. People tag pictures, videos, and other resources with a couple of keywords to easily retrieve them in a later stage. The following features of collaborative tagging are generally attributed to their success and popularity (Mathes 2004; Quintarelli 2005; Wu et al. 2006):

- *Low cognitive cost and entry barriers.* The simplicity of tagging allows any Web user to classify their favourite Web resources by using keywords that are not constrained by predefined vocabularies.
- *Immediate feedback and communication.* Tag suggestions in collaborative tagging systems provide mechanisms for users to communicate implicitly with each other through tag suggestions to describe resources on the Web.
- *Quick adaptation to changes in vocabulary.* The freedom provided by tagging allows fast response to changes in the use of language and the emergency of new words. Terms like Web2.0, ontologies and social network can be used readily by the users without the need to modify any pre-defined schemes.
- *Individual needs and formation of organization.* Tagging systems provide a convenient means for Web users to organize their favourite Web resources. Besides, as the systems develop, users are able to discover other people who are also interested in similar items.

- *Scalability*. Predefined vocabularies become imprecise when a domain grows. Instead, tags can reach a nearly unlimited granularity.
- *Serendipity*. Controlled vocabularies are designed to ease retrieval. Less popular content that resides in the so-called long-tail of the information space is hard to find. Tags enable users to discover long-tail information by browsing through the folksonomy network of items, tags, and users.
- *Inclusiveness*. The set of potential tags includes every user's views, preferences, or language as well as all potential topics.

Since tags are created by individual users in a free form, one important problem facing tagging is to identify most appropriate tags, while eliminating noise and spam. For this purpose, Noll et al. (2009) define a set of general criteria for a good tagging system.

- *High coverage of multiple facets*. A good tag combination should include multiple facets of the tagged objects. The larger the number of facets the more likely a user is able to recall the tagged content.
- *High popularity*. If a set of tags are used by a large number of people for a particular object, these tags are more likely to uniquely identify the tagged content and the more likely to be used by a new user for the given object.
- *Least-effort*. The number of tags for identifying an object should be minimized, and the number of objects identified by the tag combination should be small. As a result, a user can reach any tagged objects in a small number of steps via tag browsing.
- *Uniformity (normalization)*. Since there is no universal ontology, different people can use different terms for the same concept. We have observed two general types of divergence: those due to syntactic variance, e.g., colour, colorize, colorize, colourise; and those due to synonym, e.g., learner and pupil, which are different syntactic terms that refer to the same underlying concept. These kinds of divergence are a double-edged sword. On the one hand, they introduce noises to the system; on the other hand it can increase recall.
- *Exclusion of certain types of tags*. For example, personally tags are less likely to be shared by different users. Thus, they should be excluded from public usage. Rather than ignoring these tags, tagging system includes a feature that auto-completes tags as they are being typed by matching the prefixes of the tags entered by the user before. This not only improves the usability of the system but also enables the convergence of tags.

Another important aspect of tagging systems is how they operate. Authors in Marlow et al. (2006) explained some important dimensions of tagging systems' design that may have immediate effect on the content and effectiveness of tags generated by the system. Some of these dimensions are listed below.

7.1.1 Tagging Rights

The permission a user has to tag resources can affect the properties of an emergent folksonomy. Systems can determine who may remove a tag. Also, systems can choose the resources which users tag or specify different levels of permissions to tag. The spectrum of tagging permissions ranges from:

- **Self-tagging**—users can only tag their own contributions (e.g. Technorati[6]). In self-tagging systems, a resource can only be tagged by its creator. For instance, most blog service providers allow tagging of a post only by its owner. Being the sole annotator, a blogger therefore does not directly interact with or is influenced by other bloggers, in contrast to collaborative tagging systems. It is therefore interesting to study the possible differences between collaborative and self-tagging systems.
- **Permission-based**—users make a decision who can tag their resources. For instance, systems can specify different levels of permissions to tag (as with the friends, family, and contact distinctions in Flickr). Likewise, systems can determine who may remove a tag, whether no one (e.g., Yahoo! Podcasts), anyone (e.g., Odeo), the tag creator (e.g., Last.fm) or the resource owner (e.g., Flickr).
- **Free-for-all**—any user can tag any resource. Free-for-all systems are obviously broad, both in the size of the group of tags assigned to a resource and in the nature of the tags assigned. For instance, tags that are assigned to a photo may be radically divergent depending on whether the tagging is performed by the photographers, their friends, or strangers looking at their photos.

7.1.2 Tagging Support

One important aspect of a tagging system is the way in which users assign tags to items. They may assign arbitrary tags without prompting, they may add tags while considering those already added to a particular resource, or tags may be proposed. There are three different categories:

- **Blind tagging**—user cannot see the other tags assigned to the resource they're tagging.
- **Viewable tagging**—users can see the other tags assigned to the resource they're tagging. Implications may be overweighting certain tags that were associated with the resource first, even if they would not have arisen otherwise.
- **Suggestive tagging**—user can see recommended tags for the resource they're tagging. The suggested tags may be based on existing tags by the same user,

---

[6]http://www.technorati.com.

tags assigned to the same resource by other users or tags generated from or other sources of related tags such as automatically gathered contextual metadata, or machine-suggested tag synonyms. A suggestive system may help associate the tag usage for a resource, or in the system, much faster than a blind tagging system would. A convergent folksonomy is more likely to be generated when tagging is not blind.

7.1.3   Aggregation

The aggregation of tags around a given resource is an important consideration. The system may allow for a multiplicity of tags for the same resource which may result in duplicate tags from different users. Alternatively, many systems ask the group to collectively tag an individual resource. It is able to distinguish two models of aggregation.

- **Bag-model**—the equal tag can be assigned to a resource multiple times, like in Delicious, allowing statistics to be generated and users to see if there is agreement among taggers about the content of the resource. In the case that a bag-model is being used, the system is afforded the ability to use aggregate statistics for a given resource to present users with the collective opinions of the taggers. In addition, these data can be used to more accurately find relationships between users, tags, and resources given the added information of tag frequencies.
- **Set-model**—a tag can be applied only once to a resource, like in Flickr and Youtube. Many systems ask the group to collectively tag an individual resource, thus denying any repetition.

7.1.4   Types of Object

The implications for the nature of the resultant tags are numerous, any object that can be virtually represented can be tagged or used in a tagging system. The types of resource tagged allow us to distinguish different tagging systems. Popular systems include simple objects, like: webpages, bibliographic materials, images, videos, songs, etc. In reality, any object that can be virtually represented can be tagged or used in a tagging system. For example, there are systems that let users tag physical locations or events (e.g., Upcoming[7]).

---

[7]http://www.upcoming.yahoo.com.

7.1.5 Sources of Material

Some systems restrict the sources through architecture (e.g., Flickr), while others restrict the sources solely through social norms (e.g., CiteULike). Resources to be tagged can be supplied:

- by the participants (YouTube,[8] Flickr, Technorati, Upcoming)
- by the system (ESP Game,[9] Last.fm, Yahoo! Podcasts[10])
- open to any Web resource (Delicious, Yahoo! MyWeb2.0[11])

7.1.6 Resource Connectivity

Resources in a tagging system may be connected to each other independently of their tags. For example, Web pages may be connected via hyperlinks, or resources can be assigned to groups (e.g. photo albums in Flickr). Connectivity can be roughly categorized as: linked, grouped, or none.

7.1.7 Social Connectivity

Users of the system may be connected. Many tagging systems include social networking facilities that allow users to connect themselves to each other based on their areas of interest, educational institutions, location and so forth. Like resource connectivity, the social connectivity could be defined as linked, grouped, or none.

The term *folksonomy* defines a user-generated and distributed classification system, emerging when large communities of users collectively tag resources. Wal (2007) and Hotho et al. (2006a, b, c) are defined a folksonomy as follows.

A folksonomy is a quadruple $F: = (U; T; I; Y)$, where U, T, I are finite sets of instances of users, tags, and items and Y defines a relation, the tag assignment, between these sets, that is, $Y \subseteq U \times T \times I$.

Folksonomies became popular on the Web with social software applications such as social bookmarking, photo sharing and weblogs. A number of social tagging sites such as Delicious, Flickr, YouTube, CiteULike have become popular. Commonly cited advantages of folksonomies are their flexibility, rapid adaptability, free-for-all collaborative customisation and their serendipity (Mathes 2004). People can in general use any term as a tag without exactly understanding the meaning of

---

[8]http://www.youtube.com.

[9]http://www.espgame.org.

[10]http://podcasts.yahoo.com.

[11]http://myweb.yahoo.com.

the terms they choose. The power of folksonomies stands in the aggregation of tagged information that one is interested in. This improves social serendipity by enabling social connections and by providing social search and navigation (Quintarelli 2005). Folksonomy shows a lot of benefits (Peters and Stock 2007):

- represent an authentic use of language,
- allow multiple interpretations,
- are cheap methods of indexing,
- are the only way to index mass information on the Web,
- are sources for the development of ontologies, thesauri or classification systems,
- give the quality "control" to the masses,
- allow searching and—perhaps even better—browsing,
- recognize neologisms,
- can help to identify communities,
- are sources for collaborative recommender systems,
- make people sensitive to information indexing.

There are two types of folksonomies: broad and narrow folksonomies (Wal 2007). The *broad* folksonomy, like Delicious, has many people tagging the same object and every person can tag the object with their own tags in their own vocabulary. Thus, in theory there is a great number of tags that all refer to the same object (item), because users might independently use very distinct tags for the same content. The *narrow* folksonomy, which a tool like Flickr represents, provides benefit in tagging objects that are not easily searchable or have no other means of using text to describe or find the object.

The narrow folksonomy is done by one or a few people providing tags that the person uses to get back to that information. The tags, unlike in the broad folksonomy, are singular in nature. The same tag cannot be associated with a single object multiple times; in other words, the creator or publisher of an object is often the person who creates the first tags (unlike in broad folksonomies), and the option to tag may be even restricted to that person. After all, a much smaller number of tags for one and the same object can be identified in a narrow folksonomy. The differences between narrow and broad folksonomies from a graph perspective are depicted in Fig. 7.1, where U is the set of users, T is the set of available tags and I is the set of items. The figure also illustrates that narrow folksonomies are a special case of broad folksonomies with the constraint that each item links to exactly one user.

7.2 A Model for Tagging Activities

Social tagging systems allow their users to share their tags of particular resources. Each tag serves as a link to additional resources tagged in the same way by other users (Marlow et al. 2006). Certain resources may be linked to each other; at the

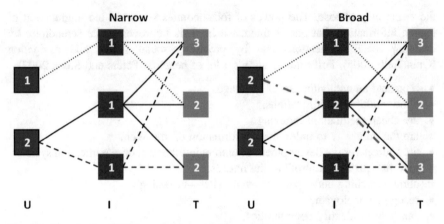

Fig. 7.1 The structural difference between narrow and broad folksonomies

same time, there may be relationships between users according to their own social interests, so the shared tags of a folksonomy come to interconnect the three groups of protagonists in social labelling systems: Users, Items, and Tags.

Many researchers (Cattuto et al. 2007; Halpin et al. 2007; Mika 2005) suggested a tripartite model that represents the tagging process:

$$Tagging : (U, T, I)$$

where U is the set of users who participate in a tagging activity, T is the set of available tags and I is the set of items being tagged. Figure 7.2 shows a conceptual model for social tagging system where users and items are connected through the tags they assign. In this model, users assign tags to a specific item; tags are

Fig. 7.2 Conceptual model of a collaborative tagging system (Marlow et al. 2006)

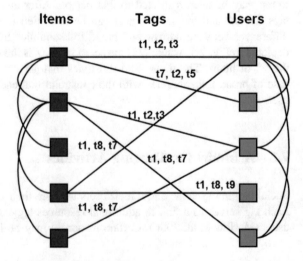

represented as typed edges connecting users and items. Items may be connected to each other (e.g., as links between Web pages) and users may be associated by a social network, or sets of affiliations (e.g., users that work for the same company).

Examination (Golder and Huberman 2006) of the collaborative tagging system, such as Delicious, has revealed a rich variety in the ways in which tags are used, regularities in user activity, tag frequencies, and great popularity in bookmarking, as well as a significant stability in the comparative proportions of tags within a given url.

- Tags may be used to identify the topic of a resource using nouns and proper nouns (i.e. photo, album or photographer).
- To classify the type of resource (i.e. book, blog, article, review, event).
- To denote the qualities and characteristics of the item (i.e. funny, useful, cool).
- A subset of tags, such as *myfavourites*, *mymusic* and *myphotos* reflect a notion of self-reference.
- Some tags are used by individuals for task organization (e.g. to read, job search, and to print).

Time is an important factor in considering collaborative tagging systems, in fact definitions and relationships among tags could vary over time. For certain users, the number of tags can become stable over time, while for others, it keeps growing. There are three hypotheses about tags behaviour over time (Halpin et al. 2007):

- **Tags convergence**: the tags assigned to a certain Web resource tend to stabilize and to become the majority. In an organizational setting with an almost stable number of contributors and a slowly growing number of topics, the number of tags used in a given domain might converge. This convergence might be interpreted as a sign of maturity of a certain topic, which in turn could trigger measures that bring the underlying tagged resources as well as the contributors, the underlying social network or community, to a higher level of maturity (Maier and Thalmann 2008).
- **Tags divergence**: tag-sets that don't converge to a smaller group of more stable tags, and where the tag distribution repeatedly changes. Xu et al. notice two types of divergence in tags: divergence due to syntactic variance: blogs, blogging, or blog and those due to synonyms cell-phone and mobile-phone. The divergence has pros and cons. It introduces noise to a tagging system, but it also can improve recall.
- **Tags periodicity**: after one group of users tag some local optimal tag-set, another group uses a divergent set but, after a period of time the new group's set becomes the new local optimal tag-set. This process may repeat and so lead to convergence after a period of instability, or it may act like a chaotic attractor.

7.3 Tag-Based Recommender Systems

Recommender systems in general recommend interesting or personalized information objects to users based on explicit or implicit ratings. Usually, recommender systems predict ratings of objects or suggest a list of new objects that the user hopefully will like the most. The approaches of profiling users with user-item rating matrix and keywords vectors are widely used in recommender systems. However, these approaches are used for describing two-dimensional relationships between users and items. In tag recommender systems the recommendations are, for a given user $u \in U$ and a given resource $r \in R$, a set $\hat{T}(u, r) \subseteq T$ of tags. In many cases, $\hat{T}(u, r)$ is computed by first generating a ranking on the set of tags according to some quality or relevance criterion, from which then the top n elements are selected as representative tags (Jaschke et al. 2007).

Personalized recommendation is used to conquer the information overload problem, and collaborative filtering recommendation is one of the most successful recommendation techniques, to date. However, collaborative filtering recommendation becomes less effective when users have multiple interests, because users have similar taste in one aspect may behave quite different in other aspects. Information got from social tagging websites not only tells what a user likes, but also why (s)he likes it. Tagging represents an action of reflection, where the tagger sums up a series of thoughts into one or more summary tags, each of which stands on its own to describe some aspect of the resource based on the tagger's experiences and beliefs (Bateman et al. 2007).

In the remainder of this section, we first describe the proposed extension with integrating tags information to improve recommendation quality. We then present two different ways for collecting tags which can work complementary to collaborative tagging, result to tag collections with improved quality.

7.3.1 Extension with Tags

The current recommender systems usually use collaborative filtering techniques, which traditionally exploit only pairs of two-dimensional data. As collaborative tagging is getting more widely used social tags as a powerful mechanism that reveals three-dimensional correlations between users–tags–items, could also be employed as background knowledge in a recommender system.

The first adaptation lies in reducing the three-dimensional folksonomy to three two-dimensional contexts: <*user, tag*> and <*item, tag*> and <*user, item*>. This can be done by augmenting the standard user-item matrix horizontally and vertically with user and item tags correspondingly (Tso-sutter et al. 2008). User tags are tags that user u uses to tag items and are viewed as items in the user-item matrix. Item tags, are tags that describe an item i, by users and play the role of users in the

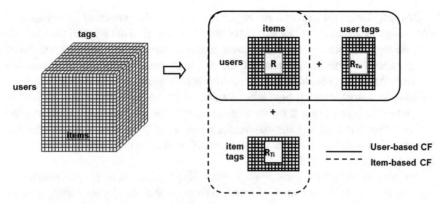

Fig. 7.3 Extend user-item matrixes by including user tags as items and item tags as users (Tso-sutter et al. 2008)

user-item matrix (see Fig. 7.3). Furthermore, instead of viewing each single tag as user or item, clustering methods can be applied to the tags such that similar tags are grouped together.

A tag based recommender system must approach several challenges to be successful in a real world application (Marinho et al. 2011):

- tags should describe the annotated item,
- items should awake the interest of the user,
- suggested items should be interesting and relevant,
- the suggestions should be traceable such that one easily understands why (s)he got the items suggested,
- the suggestions must be delivered timely without delay,
- the suggestions must be easy to access (i.e., by allowing the user to click on them or to use tab-completion when entering tags),
- the system must ensure that recommendations do not obstruct the normal usage of the system.

Recommending tags can serve various purposes, such as: increasing the chances of getting an item annotated, reminding a user what an item is about and consolidating the vocabulary across the users.

7.3.2 Collecting Tags

The quality of tags can directly affect the recommendation process. Collaborative tagging exploits the "wisdom of crowds". The following alternative ways for collecting tags can work complementary to collaborative tagging, resulting to tag collections with improved quality (Marinho et al. 2011).

Tagging based on experts: An expert is someone who possesses a high level of knowledge in a particular domain (Marinho et al. 2011). This implies that experts provide tags that are objective and cover multiple aspects. The main advantage of using experts is the high-quality of the resulting tags, especially for e-learning systems. An expert should be someone who is able to recognizes the usefulness of an item/document/learning object before the others do, thus becoming the first to assign tags to it and bring it to the attention of other users. Generally speaking, the earlier a user has tagged a document, the more credit (s)he should receive for his actions. This comes, of course, to the cost of manual work, which is both time consuming and expensive.

Tagging based on content: Several items, like URLs, songs, etc., contain a rich content. By crawling associated information from the Web and by converting it into a suitable representation, tags can be collected using data mining algorithms. In the tag recommendation task some of the tags to be predicted in the test set never appeared in the training set, which forced the participants (Tatu et al. 2008) to use the textual content of the items to come up with new tags. The advantage of content-based tags is that no humans must be directly involved during the collection process. The disadvantages are that these tags can be noisy and that their computation is intensive.

Compared to the alternative methods, social tagging has the advantage of producing large-scale tag collections. The quality of tags generally improves with a large number of taggers. Nevertheless, social tagging is prone to the cold-start problem, as new resources are seldom tagged. The main advantages and disadvantages of the described approaches are shown in Table 7.1 (Marinho et al. 2011).

In the rest of this chapter, we analyse a new approach which improves the understanding of learners, incorporating the tag information into the recommendation process. We first describe the proposed features of collaborative tagging that are generally attributed to their success in e-learning. We then present several different recommendation algorithms for developing tag-based recommender systems which are suitable for e-learning environments. The *FolkRank* algorithm, developed as a folksonomy search engine by using the graph model, is reported in Sect. 7.4.1. Probabilistic latent semantic analysis (PLSA), as a novel statistical technique for the analysis of two-mode and co-occurrence data, is described in Sect. 7.4.2. Section 7.4.3 reviews a method for tag-based profile construction with collaborative filtering based on collaborative tagging. Tensor factorization technique for tag recommendation is shown in Sect. 7.4.4. In Sect. 7.4.5 three types of "Most Popular Tags" algorithms are examined.

Table 7.1 Characterization of tag collection methods (Marinho et al. 2011)

Method	Advantages	Disadvantages
Social tagging	Scalability, social context, "wisdom of crowds"	Polysemy, cold start
Experts	Accurate tags	Costly process, difficult scalability
Content-based	Automation, avoids cold start	Noise, computationally intensive

7.4 Applying Tag-Based Recommender Systems to E-Learning Environments

Collaborative tagging systems have grown in popularity over the Web in the last years on account of their ability to categorize and recover content using open-ended tags (Godoy and Amandi 2008). The increasing number of users providing information about themselves through collaborative tagging activities caused the appearance of tag-based profiling approaches, which assume that users expose their preferences for certain contents through tag assignments. Thus, the tags could be interesting and useful information to enhance recommender system's algorithms. Tag-based recommender systems (Milicevic et al. 2010) analyse tags, discover preferences of a given user and provide suggestions for the user which items could be interesting. The main advantage of the tag-based recommenders is that user preferences and interests are expressed by used tags of the given person. Therefore, these recommenders provide more accurate and personalized recommendations. On the other hand, majority of the tag-based recommenders consider only textual (syntactical) similarities among tags. It causes problems when there are tag synonyms and according to the syntactical similarity these relations will not be revealed. The similar problem can occur when a given tag has more different meanings—so called polysemy. These issues are handled by various techniques which extend standard tag-based recommenders and provide semantically more accurate recommendations (as analysed in Sect. 7.3.1). In this chapter we investigate the suitability of tag-based recommender systems into a new context: e-learning. The innovation with respect to the e-learning systems lies in their ability to support learners in their own learning paths by recommending tags and learning items, and also their ability to promote the learning performance of individual learners (Manouselis et al. 2011).

Using tags enables useful item organization and browsing techniques, such as "pivot browsing" (Millen et al. 2006), which provides a simple and effective method for discovering new and relevant items. Learners could benefit from writing tags in two important ways: first, tagging is proven to be a meta-cognitive strategy that involves learners in active learning and engages them with more effectively in the learning process. As summarized by Bonifazi et al. (2002), tags could help learners to remember better by highlighting the most significant part of a text, could encourage learners to think when they add more ideas to what they are reading, and could help learners to clarify and make sense of the learning content while they try to reshape the information. Learners' tags could create an important trail for other learners to follow by recording their thoughts about specific tutorial resource and could give more comprehensible recommendation about the resources. While the viewing of tags used on a webpage can give a learner some idea of its importance and its content, it falls short of supporting a learner in finding the exact point of interest within the page. The following features of collaborative tagging are generally attributed to their success in e-learning (Bateman et al. 2007; Dahl and Vossen 2008; Doush et al. 2012):

1. The information provided by tags makes available insight on learner's comprehension and activity, which is useful for both educators and administrators.
2. Collaborative tagging has potential to further enhance peer interactions and peer awareness centered on learning content.
3. Tagging, by its very nature, is a reflective practice which can give learners an opportunity to summarize new ideas, while receiving peer support through viewing other learners' tags/tag suggestions.
4. In e-learning there is a lack of the social cues that inform instructors about the understanding of new concepts by their learners. Collaborative tags, created by learners to categorize learning contents, would allow instructors to reflect at different levels on their learners' progress. Tags could be examined at the individual level to examine the understanding of a learner (e.g. tags that are out of context could represent a misconception), while tags examined at the group level could identify the overall progress of the class. Working with instructors of online courses employing tagging would help to shed light on the perceived benefits of reflection based on tags.
5. Tagging provides possible solutions for learners' engagement in a number of different annotation activities—add comments, corrections, links, or shared discussion. E-learning systems currently lack sufficient support for self-organization and annotation of learning content (Bateman et al. 2007). However, walk through a university campus we can see learners engaged in a number of annotation activities. These include writing notes, creating marginalia in books, highlighting text, creating dog ears on pages or bookmarking pages. During lectures as many as 99 % of learners take notes (Palmatier and Bennett 1974), and 94 % of learners at the post-secondary level believe that note-taking is an important educational activity (Wiley 2007). In this sense tagging is beneficial to note-taking, since tags represent an aspect or cue to be used in the tagger's recall process.

Traditionally, e-learning systems intend to provide direct customized instruction to learners by finding the mismatches between the knowledge of the expert and the actions that reflect the assimilation of that knowledge by the learner (Santos and Boticario 2008). Their main limitations are:

1. e-learning systems are specific of the domain for which they have been designed (since they have to be provided with the expert knowledge) and
2. it is unrealistic to think that it is possible to code in a system all the possible responses to cover the specific needs of each learner at any situation of the course.

In this sense, a dynamic support that recommends learners what to do to achieve their learning goals is desirable. Also, such systems should have capability to find appropriate content on the Web, and capability to personalize and adjust this content based on the system's examination of its learners and the collected tags given by the learners and domain experts.

7.4.1   FolkRank Algorithm

The *FolkRank* algorithm has been inspired by the PageRank algorithm which exploits the network structures of Web pages. The PageRank algorithm assumes that a hyperlink from one page to another is a vote from the former to the latter (Page et al. 1999). The more votes a page receives, the more important that page is assumed to be.

This idea is similar to an item which is tagged with important tags by important learners becomes important itself. For example, one definition/example/task could be tagged with important tags by important learner with high knowledge level. Such definition/example/task may be considered as an important definition/example/task. The same holds, symmetrically, for tags and learners. The distribution of weights can thus be described as the fixed point of a weight passing scheme on the Web graph.

The hyperlinks indicate how important a learning object is. Tags, though, incorporate more information than does a simple hyperlink, which represents a learner created textual description of a LO. Thus, intuition would suggest that additional information can be harnessed in some way to create better search results. That tags can provide useful information for new statistical approaches which take into account human-based voting and knowledge, using algorithms similar to PageRank.

The *FolkRank* algorithm adopted the same weight spreading approaches as in the PageRank. The main difference, however, lies in the graph (Hotho et al. 2006a, b, c). In the *FolkRank*, the graph of tags has no direction, while the PageRank uses directed graphs.

Folksonomy-Adapted PageRank. The *FolkRank* algorithm transforms the hypergraph formed by the traditional tag assignments into an undirected, weighted tripartite graph $G_F = (V_F, E_F)$, which serves as input for an adaption of PageRank (Page et al. 1999). At this, the set of nodes is $V_F = L \cup T \cup I$ and the set of edges is given via $E_F = \{\{l,t\}, \{t,i\}, \{l,i\} | (l,t,i) \in Y\}$. The weight ω of each edge is determined according to its frequency within the set of tag assignments, i.e. $\omega(l,t) = |\{i \in I : (l,t,i) \in Y\}|$ is the number of items the learner l tagged with keyword t.

Accordingly, $\omega(t,i)$ counts the number of learners who annotated item i with tag t, and $\omega(l,i)$ determines the number of tags a learner l assigned to an item i. With G_F represented by the real matrix A, which is obtained from the adjacency matrix by normalizing each row to have a sum equal to 1, and starting with any vector $\vec{\omega}$ of non-negative reals, adapted PageRank iterates as $\vec{\omega} \leftarrow dA\vec{\omega} + (1-d)\vec{p}$.

Adapted *PageRank* utilizes vector $\vec{p}$, used to express learner preferences by giving a higher weight to the components which represent the learner's preferred Web pages, fulfilling the condition $\|\vec{\omega}\|_1 = \|\vec{p}\|_1$. Its influence can be adjusted by d $\in [0; 1]$. Based on this, *FolkRank* algorithm defined as follows.

The *FolkRank* algorithm computes a topic specific ranking in folksonomies: If $\vec{p}$ specifies the preference in a topic (e.g. preference for a given tag), $\vec{\omega}_0$ is the result of applying the adapted *PageRank* with d = 1 and $\vec{\omega}_1$ is the result of applying the adapted *PageRank* with some d < 1, then $\vec{\omega} = \vec{\omega}_1 - \vec{\omega}_0$ is the final weight vector. $\vec{\omega}_1[x]$ denotes the *FolkRank* of $x \in V$ (Hotho et al. 2006a, b, c).

FolkRank yields a set of related learners and items for a given tag. Following these observations, *FolkRank* can be used to generate recommendations within a folksonomy system. These recommendations can be presented to the learner at different points in the usage of a folksonomy system (Hotho et al. 2006a, b, c):

- Learning objects that are of potential interest to a learner can be suggested to him. This kind of recommendation increases the chance that a learner finds useful items that (s)he did not even know existed by "serendipitous" browsing.
- When using a certain tag, other related tags can be suggested. This can be used, for instance, to speed up the consolidation of different terminologies and thus facilitate the emergence of a common vocabulary.
- While folksonomy tools already use simple techniques for tag recommendations, *FolkRank* additionally considers the tagging behaviour of other learners.
- Other learners that work on related topics can be made explicit, improving thus the knowledge transfer within organizations and fostering the formation of communities.

FolkRank is robust against online updates since it does not need to be trained every time a new learner, item or tag enters the system. However, *FolkRank* is computationally expensive and not trivially scalable. It is more suitable for systems where real-time recommendations are not a requirement. (Hotho et al. 2006a, b, c) investigated *FolkRank* ranking in contrast to the *Adapted PageRank*. Results present that the *Adapted PageRank* ranking contains many globally frequent tags, while the *FolkRank* ranking provides more personal tags. While the differential nature of the *FolkRank* algorithm usually pushes down the globally frequent tags such as "Web", though, this happens in a distinguished manner: *FolkRank* will keep them in the top positions, if they are indeed relevant to the learner under consideration.

7.4.2 PLSA

Probabilistic latent semantic analysis (*PLSA*) is a useful statistical technique for the analysis of two-mode and co-occurrence data, which has applications in information retrieval and filtering, natural language processing, machine learning from text, and in related areas. *PLSA* has been shown to improve the quality of collaborative filtering based recommenders (Hofmann 1999) by assuming an underlying lower dimensional latent topic model.

Web users show different types of behaviour depending on their information needs and their intended tasks. These tasks are captured implicitly by a collection of actions taken by users during their visits to a site. For example, in a dynamic e-learning Web site, user tasks may be reflected by sequences of interactions with application to browse course information, to register for courses, to read a tutorial, to study an example or to solve a test. The identification of intended learner tasks can shed light on various types of learner navigational behaviours. There may be many learner groups with different (but overlapping) behaviour types. These may include learners who engage in reading content by browsing through a variety of learning objects in different categories; learners who are goal-oriented showing interest in a specific category; or learners who prefer to go through the course step by step, in a linear way with each step following logically from the previous one, or learners who tend to learn in large leaps. Most current Web usage mining systems use different data mining techniques, such as clustering, association rule mining, and sequential pattern mining to extract usage patterns from user historical navigational data (Pierrakos et al. 2003). Generally, these usage patterns are standalone patterns at the page view level. They, however, do not capture the intrinsic characteristics of Web users' activities, nor can they quantify the underlying and unobservable factors that lead to specific navigational patterns.

Thus, to better understand the factors that lead to common navigational patterns, it is necessary to develop techniques that can automatically characterize the users' underlying navigational objectives and to discover the hidden semantic relationships among users as well as between users and Web objects. A common approach for capturing the latent or hidden semantic associations among co-occurring objects is *Latent Semantic Analysis* (*LSA*) (Deerwester et al. 1990). It is mostly used in automatic indexing and information retrieval (Hofmann 1999), where *LSA* usually takes the (high dimensional) vector space representation of documents based on term frequency as a starting point and applies a dimension reducing linear projection, such as *Singular Value Decomposition* (*SVD*) to generate a reduced latent space representation (Deerwester et al. 1990).

Probabilistic latent semantic analysis (*PLSA*) models, proposed by Hofmann (1999, 2003), provide a probabilistic approach for the discovery of latent variables which is more flexible and has a more solid statistical foundation than the standard *LSA*. The basis of *PLSA* is a model often referred to as the aspect model [17]. Assuming that there exist a set of hidden factors underlying the co-occurrences among two sets of objects, *PLSA* uses Expectation-Maximization (EM) algorithm to estimate the probability values which measure the relationships between the hidden factors and the two sets of objects.

According to Hotho et al. (2006a, b, c) a folksonomy can be described as a tripartite graph whose vertex set is partitioned into three disjoint sets of users $U = \{u_1, ..., u_l\}$, tags $T = \{t_1, ..., t_n\}$ and items $I = \{i_1, ..., i_m\}$. This model can be simplified to two bipartite models. The collaborative filtering model IU is built from the item user co-occurrence counts $f(i, u)$. The annotation-based model IT derives from the co-occurrence counts between items and tags $f(i, t)$. In the case of social bookmarking IU becomes a binary matrix $(f(i, u) \in \{0, 1\})$, as users can bookmark a given Web resource only once.

The aspect model of *PLSA* associates the co-occurrence of observations with a hidden topic variable $Z = \{z_1 \dots z_k\}$. In the context of collaborative filtering an observation corresponds to the bookmarking of an item by a user and all observations are given by the co-occurrence matrix IU (Wetzker et al. 2009). Users and items are assumed independent given the topic variable Z. The probability that an item was bookmarked by a given user can be computed by summing over all latent variables Z:

$$P(i_m|u_l) = \sum_k P(i_m|z_k)P(z_k|u_l),$$

Analogous to (3), the conditional probability between tags and items can be defined as:

$$P(i_m|t_n) = \sum_k P(i_m|z_k)P(z_k|t_n),$$

Following the Cohn's and Hofmann's procedure (2001), we can now combine both models based on the common factor $P(i_m|z_k)$ by maximizing the log-likelihood function:

$$L = \sum_m \left[\alpha \sum_l f(i_m, u_l) \log P(i_m|u_l) + (1 - \alpha) \sum_n f(i_m, t_n) \log P(i_m|t_n) \right]$$

where α is a predefined weight for the influence of each two-mode model. Using the *Expectation-Maximization* (*EM*) algorithm (Cohn and Hofmann 2001) it can be performed maximum likelihood parameter estimation for the aspect model. The standard procedure for maximum likelihood estimation in latent variable models is the *Expectation Maximization* (*EM*) algorithm (Arenas-García et al. 2007). *EM* alternates two coupled steps:

1. an *expectation* (*E*) step where posterior probabilities are computed for the latent variables,
2. a *maximization* (*M*) step, where parameters are updated. Standard calculations yield the E-step equation:

$$P(z_k|u_l, i_m) = \frac{P(i_m|z_k)P(z_k|u_l)}{P(i_m|u_l)}$$

$$P(z_k|t_n, i_m) = \frac{P(i_m|z_k)P(z_k|t_n)}{P(i_m|t_n)}$$

and then re-estimate parameters in the maximization (*M*) step as follows:

$$P(z_k|u_l) \propto \sum_m f(u_l, i_m)P(z_k|u_l, i_m)$$

$$P(z_k|t_n) \propto \sum_m f(t_n, i_m)P(z_k|t_n, i_m)$$

$$P(i_m|z_k) \propto \alpha \sum_l f(u_l, i_m)P(z_k|u_l, i_m)$$

$$+ (1 - \alpha) \sum_n f(t_n, i_m)P(z_k|t_n, i_m)$$

Based on the iterative computation of the above *E* and *M* steps, the *EM* algorithm monotonically increases the likelihood of the combined model on the observed data. Using the parameter, this model can be easily reduced to a collaborative filtering or annotation-based model by setting to 1.0 or 0.0 respectively.

It is possible to recommend items to a user u_l weighted by the probability $P(i_m|u_l)$. For items already bookmarked by the user in the training data this weight set to 0, thus they are appended to the end of the recommended item list.

PLSA as a hybrid approach to the task of item recommendations in folksonomies that includes user generated annotations produces better results than a standard collaborative filtering or annotation-based methods.

7.4.3   *Collaborative Filtering Based on Collaborative Tagging*

Collaborative filtering is based on the assumption that people with similar tastes (i.e., people who agreed in the past) will prefer similar items (i.e., will agree in the future) (Shardanand and Maes 1995). Traditionally, *collaborative filtering* techniques predict ratings of items or suggest a list of new items that the user will like the most. In the case of e-learning, collaborative systems track past actions of a group of learners to make a recommendation for individual members of the group (Tan et al. 2008). Based on the assumption that learners with similar past behaviors (browsing, learning path, item ratings or grades that they received by the system) have similar interests and similar appropriate level of knowledge, a collaborative filtering system recommends learning objects of the given learner. This approach relies on a historic record of all learner interests such as can be inferred from their ratings of the items (learning objects/learning actions) on a website. Rating can be explicit (explicit ratings or learner satisfaction questionnaires) or implicit (from the studying patterns or click-stream behaviour of the learners).

The learner profiles can be represented in a learner-item matrix $X \in R^{m \times n}$, for m learners and n items. The matrix can be decomposed into row vectors:

$$X := [x_1, \ldots, x_m]^T \ with \ x_l := [x_{l,1}, \ldots, x_{l,n}], \quad for \ l := 1, \ldots, m$$

where $x_{l,i}$ indicates that learner l rated item i by $x_{l,i} \in R$. Each row vector x_l corresponds thus to a learner profile representing the item's ratings of a particular learner. This decomposition usually leads to algorithms that leverage learner-learner similarities, such as the well-known user-based collaborative filtering (Resnick et al. 1994). Given two learners x_u and x_v, we then quantify learners' similarity $sim(x_u, x_v)$ as the cosine of the angle between their vectors:

$$sim(x_u, x_v) = \frac{\langle x_u, x_v \rangle}{\|x_u\| \|x_v\|}.$$

Alternatively, *Pearson Correlation* (and its variations—e.g., weighted Pearson) (Herlocker et al. 2004) could be used.

The matrix X can alternatively be represented by its column vectors:

$$X := [x_1, \ldots, x_n]^T \ with \ x_r := [x_{u,1}, \ldots, x_{m,r}], \quad for \ u := 1, \ldots, n$$

in which each column vector x_r corresponds to a specific item's ratings by all m learners. This representation usually leverages item-item similarities and leads to item-based CF algorithms (Deshpande and Karypis 2004).

Collaborative filtering for tag recommendations in folksonomies aim at modelling user interests based on their historical tagging behaviours, and recommend tags to a user from similar users or user groups (Golder and Huberman 2006). Tags are used for navigation, finding resources and serendipitous browsing and thus provide an immediate benefit for users. CF tag-based RSs usually include tag recommendation mechanisms easing the process of finding good tags for a resource, but also consolidating the tag vocabulary across users. Specifically, during the collaborative step, users who share similar tagging behaviours with the user to whom we want recommend tags too are chosen based on the between-user similarities, which are calculated based on the users' tagging history. This step usually requires a pre-computed look-up table for the between-user similarities, which is usually in the form of weighted symmetric matrices.

In the case of e-learning, collaborative tags represent a form of practical metadata, which could be useful for detailed learning object descriptions. Also, tagging provides possible solutions for learners' engagement in a number of different annotation activities—add comments, corrections, links, or shared discussion. Learners' tags could create an important trail for other learners to follow by recording their thoughts about the specific resources and could give more comprehensible recommendation about the resources. Therefore, we can conclude that tag collection of like-minded learners offer active learners advice on what is

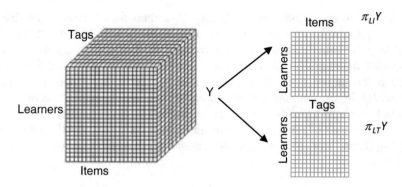

Fig. 7.4 Projections of Y into the learner's item and learner's tag spaces (Tso-sutter et al. 2008)

important in a tutorial, what is difficult in a lesson, which example is useful etc. (e.g. learners could observe tag clouds describing course concepts).

Traditionally, collaborative filtering techniques exploit only pairs of two-dimensional data. Thus, because of the ternary relational nature of folksonomy which provides a 3-dimensional relationship between users, items and tags, traditional CF cannot be applied directly in e-learning systems. The first adaptation lies in reducing the ternary relation Y to a lower dimensional space (Marinho and Schmidt-Thieme 2008). In the case of user-based CF, we consider matrix X as alternatively the two 2-dimensional projections (Fig. 7.4) for learner l, and item i:

$$\pi_{LI}^Y \in \{0,1\}^{|L|\times|I|} \ with \left(\pi_{LI}^Y\right)_{l,i}: = 1 \quad \textit{if there exist } t \in T \ s.t \ (L,t,i) \in Y \textit{ and } 0 \textit{ else}$$

$$\pi_{LT}^Y \in \{0,1\}^{|L|\times|T|} \ with \left(\pi_{LT}^Y\right)_{l,t}: = 1 \quad \textit{if there exist } i \in L \ s.t \ (L,t,i) \in Y \textit{ and } 0 \textit{ else}$$

We first compute the set N_l^k of the k learners that are most similar to learner l, based on the row decomposed version of X and for a given k:

$$N_l^k := \arg\max_{v\in L\setminus\{l\}}^{k} sim(x_u, x_v)$$

where the superscript in the *argmax* function indicates the number $k \in N$ of neighbors to be returned. Having the neighbourhood determined, we can extract the set $\hat{T}(I, i)$ of s recommended tags for a given user l, a given item i, and some $s \in N$, as follows:

$$\hat{T}(I, i) := \arg\max_{t\in T}^{s} \sum_{v\in N_l^k} sim(x_u, x_v)\delta(v, t, i)$$

$$\textit{where } \delta(v,t,i) := 1 \textit{ if } \delta(v,t,i) \in Y \textit{ and } 0 \textit{ else}$$

In order to apply collaborative filtering algorithms for tag recommendation in folksonomies, some data transformation must be performed. Such transformations lead to information loss, which can lower the recommendation quality, but collaborative filtering algorithms are robust against online updates since it does not need to be trained every time a new learner, item or tag enters the system. Especially in the learning process, consideration of like-minded learners that worked on related topics can be of great importance for active learner. Furthermore, as Sood et al. (2007) point out, tag recommendations "fundamentally change the tagging process from generation to recognition" which requires less cognitive effort and time.

7.4.4 Tensor Factorization Technique for Tag Recommendation

Most developing recommendation algorithms (Hotho et al. 2006a, b, c; Xu et al. 2006) try to exploit the provided data (users—u, items—i, tags—t) only in 2-dimensional relations. These pairs: (users, tags), (users, items), (tags, items) are analysed by the different types of the algorithms which determine the most relevant and appropriate content—tags or items for the users. However, these algorithms do not consider the 3 dimensions of the problem altogether, and therefore they miss a part of the semantics that is carried by the 3-dimensions.

Researcher Symeonidis et al. (2008) recognized that involving and exploring existent relationships between tags, users and items can reveal more relevant effects. They suggested tensor based technique which can address the problem of recommendation by capturing the multimodal perception of items by particular users (learning materials by particular learners). It can perform 3-dimensional analysis on the social tags data, attempting to discover the latent factors that determine the associations among the triplets user–tag–item. Consequently, items can be recommended according to the captured associations. That is, given a learner and a tag, the purpose is to predict whether and how much the learner is likely to label with this tag a specific learning item.

As a simple example, let us consider the social tagging system of learners in e-learning system, we developed—Protus 2.1. Assume we have two learners. One would like to revise (study/repeat) the examples of examination task and therefore has tagged Example 4 as "useful" and Example 6 as "suitable". Another learner learned studiously and has tagged introductory example as "useful" and "basics" for learning next, complex learning material. When wanting to study "useful" examples, both learners are recommended some examples, while the first learner is expecting the examples of examination task and the other prefers the introductory examples.

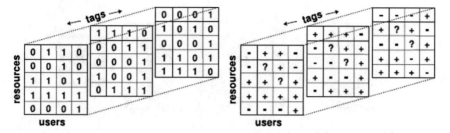

Fig. 7.5 Tensor representations—*Left* (Symeonidis et al. 2008), *Right* (Rendle et al. 2009)

Recommendation algorithms based on tensor factorization generate their recommendations using ranking score which is computed according to spectral attributes extracted from the underlying folksonomy data structure. By representing $\mathcal{Y}$ as a tensor, one is able to exploit the underlying latent semantic structure in $\mathcal{A}$ formed by multi-way correlations between users, tags, and items. There are different ways to represent $\mathcal{Y}$ as $\mathcal{A}$ (Symeonidis et al. 2008), for example, proposed to interpret $\mathcal{Y}$ as a sparse tensor (Fig. 7.5 left) in which 1 indicates positive feedback and the remaining data as 0:

$$a_{u,t,i} = \begin{cases} 1, & (u,t,i) \in Y \\ 0, & else \end{cases}$$

Rendle et al. (2009) on the other hand, distinguish between positive and negative examples and missing values in order to learn personalized ranking of tags. The idea is that positive and negative examples are only generated from observed tag assignments. All other entries, i.e., all tags for an item that a user has not tagged yet, are assumed to be missing values (Fig. 7.5 right).

In this section, we will analyse the recommendation systems based on tensor factorization using Higher Order Singular Value Decomposition (*HOSVD*). We first provide an outline of SVD approach (Singular Value Decomposition), tensor and Higher Order Singular Value Decomposition (*HOSVD*) method. Next, we analyse the steps of the Ranking with Tensor Factorization (*RTF*) algorithm. In the rest of the section, we denote tensors by calligraphic uppercase letters (e.g., $\mathcal{A}, \mathcal{B}$), matrices by uppercase letters (e.g., A, B), scalars by lowercase letters (e.g., a, b), and vectors by bold lowercase letters (e.g., **a**, **b**).

7.4.4.1   SVD Algorithm

The tensor reduction technique based on a SVD (Berry et al. 1995) calculates matrix approximation. The SVD of a matrix $F_{I_1 \times I_2}$ can be written as a product of three matrices:

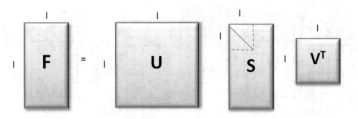

Fig. 7.6 Visualization of the matrix SVD

$$F_{I_1 \times I_2} = U_{I_1 \times I_1} \cdot S_{I_1 \times I_2} \cdot V^{\mathrm{T}}_{I_2 \times I_2}$$

where U is the matrix with the left singular vectors of F, V^{T} is the transpose of the matrix V with the right singular vectors of F and S is the diagonal matrix of singular values of F. Visualization of the matrix SVD is shown in Fig. 7.6.

By preserving only the largest $c < \{I_1, I_2\}$ singular values of S, SVD results to matrix $\hat{F}$, which is an approximation of F. The tuning of c is empirically determined by the information percentage that is preserved compared to the original matrix (De Lathauwer et al. 2000).

7.4.4.2 Tensors and HOSVD Algorithm

A tensor is a multi-dimensional matrix. N-order tensor $\mathcal{A}$ is denoted as $\mathcal{A} \in R^{I_1 \times I_2 \times ... \times I_N}$, with elements $a_{i1,...,iN}$.

Definition. The n-mode product of a tensor $\mathcal{A} \in R^{I_1 \times I_2 \times ... \times I_N}$ by a matrix $U \in R^{I_n \times I_n}$, denoted by $\mathcal{A} \times_n U$, is an $(I_1 \times I_2 \times ... \times I_{n-1} \times J_n \times I_{n+1} \times ... \times I_N)$—tensor of which the entries are given by De Lathauwer (1997):

$$(\mathcal{A} \times_n U)_{i_1 i_2 ... i_{n-1} j_n i_{n+1} ... i_N} = \sum_{i_n} a_{i_1 i_2 ... i_{n-1} i_n i_{n+1} ... i_N} u_{j_n i_n}$$

We only use 3-order tensors (the three dimensions are: u-users, i-items and t-tags) where $\mathcal{A} \in R^{u \times t \times i}$. Each tensor element measures the preference of a (user u, tag t) pair on an item i. Tensor A can be metricized i.e., represented by building matrix representations in which all the column (row) vectors are stacked one after the other.

Thus, after the unfolding of tensor A for all three modes, we create 3 new matrices A_1, A_2 and A_3 as follows (De Lathauwer et al. 2000):

$$A_1 \in R^{I_u \times I_t I_i},$$
$$A_2 \in R^{I_t \times I_u I_i},$$
$$A_3 \in R^{I_u I_t \times I_i}$$

Fig. 7.7 Visualization of the three unfoldings of a 3-order tensor A

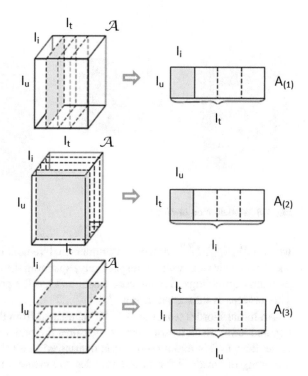

where A_1, A_2, and A_3 are called the 1-mode 2-mode, 3-mode matrix unfolding of $\mathcal{A}$, respectively. The unfolding of $\mathcal{A}$ *in* the three modes, is illustrated in Fig. 7.7.

In terms of n-mode products, SVD on a regular two-dimensional matrix (i.e. 2-order tensor), can be rewritten as follows (De Lathauwer et al. 2000):

$$F = S \times_1 U^{(1)} \times_2 U^{(2)}$$

where $U^{(1)} = \left(U_1^{(1)} U_2^{(1)} \ldots U_{I_u}^{(1)} \right)$ is a unitary $(I_u \times I_u)$-matrix, $U^{(2)} = \left(U_1^{(2)} U_2^{(2)} \ldots U_{I_u}^{(2)} \right)$ is a unitary $(I_t \times I_t)$-matrix and S is an $(I_u \times I_t)$-matrix with the following properties:

- Pseudodiagonality $\left(S = diag\left(\sigma_1, \sigma_2, \ldots, \sigma_{\min\{I_u, I_t\}} \right) \right)$
- Ordering $\left(\sigma_1 \geq \sigma_2 \geq \cdots \geq \sigma_{\min\{I_u, I_t\}} \geq 0 \right)$

By extending this form of SVD, the *HOSVD* of 3-order tensor $\mathcal{A}$ can be written as follows (De Lathauwer et al. 2000):

$$\mathcal{A} = S \times_1 U^{(1)} \times_2 U^{(2)} \times_3 U^{(3)}$$

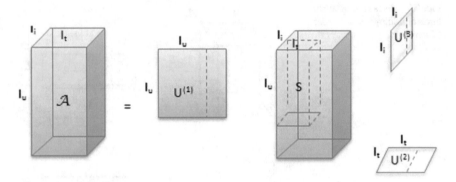

Fig. 7.8 Visualization of *HOSVD*

where $U^{(1)}$, $U^{(2)}$, $U^{(3)}$ contain the orthonormal vectors (called the 1-mode 2-mode and 3-mode singular vectors, respectively) spanning the column space of the A_1, A_2, A_3 matrix unfoldings. S is the core tensor and has the property of all orthogonality. This process is illustrated in Fig. 7.8.

An initial 3-order tensor $\mathcal{A} \in R^{u \times t \times i}$ is created from the usage data triplets (user, tag, and item). Each tensor element measures the preference of a (user u, tag t) pair on an item i. This initial tensor $\mathcal{A}$ is matricized in all three modes. Thus, after the unfolding of tensor $\mathcal{A}$ for all three modes, we create 3 new matrices A_1, A_2, A_3, as follows.

$$A_1 \in R^{I_u \times I_t I_i},$$
$$A_2 \in R^{I_t \times I_u I_i},$$
$$A_3 \in R^{I_u I_t \times I_i}$$

SVD is applied on these three matrix unfoldings. It results into total 9 new matrices.

$$A_1 = U^{(1)} \cdot S_1 \cdot V_1^T$$
$$A_2 = U^{(2)} \cdot S_2 \cdot V_2^T$$
$$A_3 = U^{(3)} \cdot S_3 \cdot V_3^T$$

For tensor dimensionality reduction, there are three parameters to be determined. The numbers c_1, c_2, and c_3 of left singular vectors of matrices $U^{(1)}$, $U^{(2)}$, $U^{(3)}$ which are determinative for the final dimension of the core tensor S. Since each of the three diagonal singular matrices S_1, S_2 and S_3, are calculated by applying SVD on matrices A_1, A_2, and A_3, respectively, we use different c_1, c_2, and c_3 values for each matrix $U^{(1)}$, $U^{(2)}$, $U^{(3)}$. The numbers c_1, c_2, and c_3 are empirically chosen by maintaining a percentage of information of the original S_1, S_2 and S_3 matrices after appropriate modification. Usually the percentage is set to 70 % of the original matrix.

The core tensor S governs the interactions among user, item and tag entities. Since we have selected the dimensions of $U^{(1)}$, $U^{(2)}$, $U^{(3)}$ matrices, we proceed to the construction of the core tensor S, as follows (Symeonidis et al. 2008):

$$S = A \times_1 U_{c_1}^{(1)^T} \times_2 U_{c_2}^{(2)^T} \times_3 U_{c_3}^{(3)^T}$$

where A is the initial tensor, $U_{c_1}^{(1)^T}$ is the transpose of the c_1-dimensionally reduced $U^{(1)}$ matrix, $U_{c_2}^{(2)^T}$ is the transpose of the c_2-dimensionally reduced $U^{(2)}$, and $U_{c_3}^{(3)^T}$ is the transpose of the c_3-dimensionally reduced $U^{(3)}$. Finally, tensor $\hat{A}$ is built by the product of the core tensor S and the mode products of the three matrices $U^{(1)}$, $U^{(2)}$, $U^{(3)}$ as follows:

$$\hat{A} = S \times_1 U_{c_1}^{(1)} \times_2 U_{c_2}^{(2)} \times_3 U_{c_3}^{(3)}$$

where S is the c_1, c_2, and c_3 reduced core tensor, $U_{c_1}^{(1)}$ is the c_1-dimensionally reduced $U^{(1)}$ matrix, $U_{c_2}^{(2)}$ is the c_2-dimensionally reduced $U^{(2)}$ matrix, $U_{c_3}^{(3)}$ is the c_3-dimensionally reduced $U^{(3)}$ matrix.

The reconstructed tensor $\hat{A}$ measures the associations among the users, tags and items. The model parameters to be learned are then the quadruple $\hat{\theta} := \left(S, U_{c_1}^{(1)}, U_{c_2}^{(2)}, U_{c_3}^{(3)} \right)$.

The basic idea is to minimize an element-wise loss on the elements of $\hat{A}$ by optimizing the square loss, i.e.,

$$\underset{\hat{\theta}}{\arg\min} \sum_{(u,t,i) \in U \times T \times I} \left(\hat{a}_{u,t,i} - a_{u,t,i} \right)^2$$

7.4.4.3 Ranking with Tensor Factorization

Rendle et al. (2009) propose Ranking with Tensor Factorization (RTF), a method for learning an optimal factorization of A for the specific problem of tag recommendations. First, the observed tag assignments are divided in positive and negative. All other entries (e.g. all tags for an item that a user has not tagged yet) are assumed to be missing values, as described in Sect. 5.3.4 (see right-hand side of Fig. 7.8). Let $P_A := \{(u,i)|\exists t \in T : (u,t,i) \in Y\}$ be the set of all distinct user/item combinations in Y, the sets of positive and negative tags of a particular $(u,i) \in P_A$ are then defined as:

$$T_{u,i}^+ := \{t|(u,i) \in P_A \wedge (u,t,i) \in Y\}$$
$$T_{u,i}^- := \{t|(u,i) \in P_A \wedge (u,t,i) \notin Y\}$$

From this, pairwise tag ranking constraints can be defined for the values of $\widehat{\mathcal{A}}$:

$$a_{u,t_1,i} > a_{u,t_2,i} \Leftrightarrow (u, t_1, i) \in T_{u,i}^+ \wedge (u, t_2, i) \in T_{u,i}^-$$

From a semantically point of view this scheme makes more sense as the user/item combinations that have no tags are the ones that the recommender system will have to predict in the future. Thus, instead of minimizing the least-squares as in the *HOSVD*-based methods, an optimization criterion that maximizes the ranking statistic AUC (area under the ROC-curve) is proposed. The AUC measure for a particular $(u, i) \in P_A$ is defined as:

$$AUC(\hat{\theta}, u, i) := \frac{1}{\left|T_{u,i}^+\right|\left|T_{u,i}^-\right|} \sum_{t^+ \in T_{u,i}^+} \sum_{t^- \in T_{u,i}^-} H_{0,5}\left(\hat{a}_{u,t^+,i} - \hat{a}_{u,t^-,i}\right)$$

where H_α is the Heaviside function:

$$H_\alpha := \begin{cases} 0, & x < 0 \\ \alpha, & x = 0 \\ 1, & x > 0 \end{cases}$$

The overall optimization task with respect to the ranking statistic AUC and the observed data is then:

$$\underset{\hat{\theta}}{\arg\max} \sum_{(u,i) \in P_A} AUC(\hat{\theta}, u, i)$$

7.4.4.4 Multi-mode Recommendations

Once $\widehat{\mathcal{A}}$ is computed, the recommendation list with the N highest scoring tags for a given user u and a given item i can be calculated by:

$$Top(u, i.N) := \underset{t \in T}{\arg\max}^N \hat{a}_{u,t,i}$$

Recommending N items to a given user u for a particular tag t can be determined in a similar manner. Moreover, other tags can be recommended to a particular user u given a specific tag t, according to the total score that results by aggregating all items that are tagged with tag t by user u. Thus, according to the data representation, tensor modeling permits multi-mode recommendations in an easy way.

To exemplify this approach, we apply the *RTF* algorithm to an illustrative example, which is illustrated in Fig. 7.9. As it can be seen, 4 learners tagged 4 different items. In the figure, the arrow lines and the numbers assigned to them represent the correspondence between the three types of entities. For example,

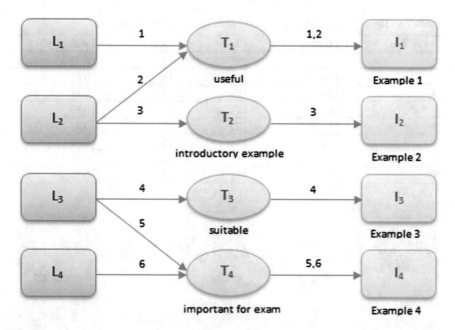

Fig. 7.9 Illustrative example

learner L_1 tagged with tag "useful" (denoted as T_1) the item "Introductory example" (denoted as I_1). From Fig. 7.9, we can see that learners L_1 and L_2 have common interests on introductory example, while learners L_3 and L_4 have common interests in the examples of examination task. A 3-order tensor $A \in R^{4 \times 4 \times 4}$ can be created from these illustrative example. We use the co-occurrence frequency of learner, tag and item as the elements of tensor A, which are given in Table 7.2.

After performing the tensor reduction analysis, we get the reconstructed tensor $\hat{A}$. Table 7.3 gives the output of the tensor reduction algorithm, which is also illustrated in Fig. 7.10. We can notice that the algorithm outputs new associations between the considered entities (the last rows in Table 7.3 and the dotted lines in Fig. 7.10). Even though in the original data, learner L_1 did not tag item I_2, the algorithm is capable to conclude that if L_1 would tag them, then L_1 would likely (likelihood 0.35) use tag "introductory example". As well, the algorithm can assume that if L_4 would tag item I_4 with another tag, then L_4 would likely (likelihood 0.44) use the tag "suitable".

	Arrow line	Learner	Tag	Item	Weight
Table 7.2 Tensor created from the used data	1	L_1	T_1	I_1	1
	2	L_2	T_1	I_1	1
	3	L_2	T_2	I_2	1
	4	L_3	T_3	I_3	1
	5	L_3	T_4	I_4	1
	6	L_4	T_4	I_4	1

Table 7.3 Tensor constructed from the usage data of the illustrative example

Arrow line	Learner	Tag	Item	Weight
1	L_1	T_1	I_1	0.72
2	L_2	T_1	I_1	0.5
3	L_2	T_2	I_2	1.18
4	L_3	T_3	I_3	0.35
5	L_3	T_4	I_4	0.35
6	L_4	T_4	I_4	0.44
7	L_1	T_2	I_2	1.18
8	L_4	T_3	I_4	0.72

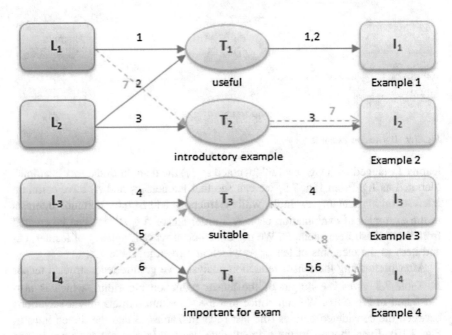

Fig. 7.10 Presentation of the tensor reduction algorithm output for the illustrative example

The tensor reduction approach is able to capture the latent associations among the multi-type data entities: learners, tags and items. These associations can further be used to improve the recommendation procedure.

7.4.5 Most Popular Tags

The web algorithms Most Popular Tags are based on tag calculations. In the rest of this section it will be presented that these methods are efficient and cheap and their computational complexity is low and therefore might be good candidates for online computation of recommendations.

If we want to compute, for a given pair (u, i), the most popular tags of the user u (or the item i), we need to linearly scan Y to calculate the occurrence counts for u's tags (or i's tags) and afterwards sort the tags we gathered by their count.

For an user $u \in U$, the set of all his tag assignments is $Y_u := Y \cap (\{u\} \times T \times I)$. The sets Y_i (for any item $i \in I$) and Y_t (for any tag $t \in T$) are defined accordingly. Similarly, for $t \in T$ and $i \in I$, $Y_{t,u} : Y \cap (\{u\} \times \{t\} \times R)$ and $Y_{t,i}$ are defined, accordingly. Finally, for a user $u \in U$, the set of all his tags can be defined as $T_u : \{t \in T | \exists i \in I : (u, t, i) \in Y\}$. The set T_i (for any item $i \in I$) is defined accordingly.

There are three types of "Most Popular Tags" algorithms:

1. Recommending the most popular tags of the folksonomy is the simplest approach. It recommends, for any user $u \in U$ and any item $i \in I$, the same set:

$$\hat{T}(u, i) := \arg\max_{t \in T}^{n} (|Y_t|)$$

2. Tags that globally are most specific to the item will be recommended when using the most popular tags by item:

$$\hat{T}(u, i) := \arg\max_{t \in T}^{n} (|Y_{t,i}|)$$

3. Since users might have specific interests for which they already tagged several items, using the most popular tags by user is another option:

$$\hat{T}(u, i) := \arg\max_{t \in T}^{n} (|Y_{t,u}|)$$

None of the aforementioned methods applied alone will, in general, provide the best recommendations. Nevertheless, the simplicity and cost efficiency of algorithms based on tag counts make them a favoured approach for use in existing folksonomy systems. Jäschke et al. (2007) experimented with a mix of the recommendations generated by variants 2 and 3 which are called most popular tags mix.

7.4.5.1 Mix of "Most Popular Tags" Recommenders

The main idea of this approach is to recommend a mix of the most popular tags of the user with the most popular tags of the item. The simplest way to mix the tags is to summate their counts and then sort them by newly obtained count:

$$\hat{T}(u,i) := \arg\max_{t \in T} \left(\left| Y_{t,u}^{n} \right| + \left| Y_{t,i} \right| \right)$$

This way of mixing is called most popular tags mix 1:1, since we just add unchanged, existing counts. For instance, if the item has been tagged three times with "popular" by other users and the user has used the tag "popular" four times for other items, the tag "popular" would get a count of seven.

Although this method already contributes good results, the influence of the user-based recommendation will be very small compared to the item-based rec-ommendation if many people have tagged this item. On the contrary, if a user has tagged many items, his most popular tags might have counts that are much higher than the counts provided by the items. Therefore, Jäschke et al. (2007) introduced another mix variant, where the tag counts of the two participating sets are nor-malized and weighted before they are added. Normalization function is defined for each tag $t \in T_i$:

$$norm_i(t) := \frac{\left| Y_{t,i} \right| - \min_{t' \in T} \left| Y_{t',i} \right|}{\max_{t' \in T} \left| Y_{t',i} \right| - \min_{t' \in T} \left| Y_{t',i} \right|}$$

For $t \in T_u$, the normalization $norm_u(t)$ is defined in analogue manner. After normalization the weights of all tags in T_i and T_u lie between zero and one—with the most popular tag(s) having weight 1 and the least important tag(s) having weight 0. A pre-defined factor $\rho \in [0,1]$ allows to balance the influence of the user and the item:

$$\hat{T}(u,i) : \arg\max_{t \in T}^{n} (\rho norm_i(t) + (1 - \rho) norm_u(t))$$

This method is called *The Most Popular Tags ρ—Mix*. T*he Most Popular Tags 0 —Mix* is just the most popular tags by user strategy, since the normalization does not change the order of the tags. Similarly, *The Most Popular Tags 1—Mix* is just the most popular tags by item strategy. However, due to normalization *The Most Popular Tags 0.5—Mix* is not identical to *The Most Popular Tags Mix 1:1*.

7.5   Limitations of Current Folksonomy and Possible Solutions

Tagging systems have the potential to improve search, recommendation and per-sonal organization while introducing new modalities of social communication. As described in this section, there has been a lot of research done on tag-based recommendation techniques that have significantly advanced the state-of-the-art in comparison to early recommender systems utilized collaborative and content-based heuristics. Despite the rapid expansion of applications that support tagging of items, the simplicity and ease of use of tagging however, lead to problems with current folksonomy systems, which hinder the growth or affect the usefulness of the systems. The problems can be classified in some categories (Gordon-Murnane 2006; Mathes 2004; Pluzhenskaia 2006; Shepitsen et al. 2008). We consider set of limitations which can directly affect the tag-based recommendation process in e-learning environments.

1. Tags have little semantics and many variations. Thus, even if a tagging activity can be considered as the learner's cognitive process, the resulting set of tags does not always correctly and consistently represent the learner's mental model.
2. As an uncontrolled vocabulary that is shared across an entire system, the terms in a folksonomy have inherent ambiguity, as different learners apply terms to items in different ways. Tag ambiguity, in which a single tag has many meanings, can falsely give the impression that items are similar when they are in fact unrelated.
3. Tag redundancy, in which several tags have the same meaning, can obfuscate the similarity among items. Redundant tags can hinder algorithms that depend on identifying similarities between items.
4. The use of different word forms such as plurals and parts of speech also worsen the problem.

There are some different approaches aiming to solve the mentioned problems. First one tries to educate learners to improve "tag literacy" (Guy and Tonkin 2006). An important condition for this way of resolving problems is to better examine learner researches about folksonomies (Bar-Ilan et al. 2006; Lin et al. 2006; Winget 2006), concerning the "deep nature" of tags (Veres 2006a), discussing aspects of the folksonomy interoperability (Veres 2006b) and the "semiotic dynamics" of folksonomies in terms of tag co-occurrences (Cattuto et al. 2007). For training the learner's selection of "good" tags it may be useful that the system would suggest some tags (MacLaurin 2010). Tag-suggestions can operate on a syntactical level (e.g., a learner attaches "graph" and the system suggests "graphics") or even on a relational level (e.g., a learner attaches "graphics" and the system suggests "image", because both words do often co-occur in items' tag clouds (Xu et al. 2006). Also, tag-suggestion can be based on experts' opinions, providing higher quality of the resulting tags that are objective and cover multiple aspects.

These extensions open huge number of opportunities for future work in this area. They can improve tag-based recommendation capabilities and make collaborative tagging systems applicable to even broader range of applications.

References

Arenas-García, J., Meng, A., Petersen, K. B., Lehn-Schioler, T., Hansen, L. K., & Larsen, J. (2007). Unveiling music structure via plsa similarity fusion. In *IEEE Workshop on Machine Learning for Signal Processing, 2007* (pp. 419–424).

Bar-Ilan, J., Shoham, S., Idan, A., Miller, Y., & Shachak, A. (2006). Structured vs. unstructured tagging–A case study. In *Proceedings of the WWW 2006 Collaborative Web Tagging Workshop.*

Bateman, S., Brooks, C., McCalla, G., & Brusilovsky, P. (2007). Applying collaborative tagging to e-learning. *WWW*, 1–7. http://doi.org/10.1.1.64.8892

Berry, M. W., Dumais, S. T., & O'Brien, G. W. (1995). Using linear algebra for intelligent information retrieval. *SIAM Review, 37*(4), 573–595. http://doi.org/10.1137/1037127

Bonifazi, F., Levialdi, S., Rizzo, P., & Trinchese, R. (2002). A web-based annotation tool supporting e-learning. In *Proceedings of the Working Conference on Advanced Visual Interfaces—AVI '02*, (p. 123). http://doi.org/10.1145/1556262.1556281

Cattuto, C., Schmitz, C., Baldassarri, A., Servedio, V. D. P., Loreto, V., Hotho, A., ... Stumme, G. (2007). Network properties of folksonomies. *Ai Communications, 20*(4), 245–262.

Cohn, D., & Hofmann, T. (2001). The missing link-a probabilistic model of document content and hypertext connectivity. *Advances in Neural Information Processing Systems*, 430–436.

Dahl, D., & Vossen, G. (2008). Evolution of learning folksonomies: Social tagging in e-learning repositories. *International Journal of Technology Enhanced Learning, 1*(1–2), 35–46.

De Lathauwer, L. (1997). *Signal processing based on multilinear algebra.* Katholieke Universiteit Leuven.

De Lathauwer, L., De Moor, B., & Vandewalle, J. (2000). A multilinear singular value decomposition. *SIAM Journal on Matrix Analysis and Applications, 21*(4), 1253–1278.

Deerwester, S., Dumais, S. T., Furnas, G. W., Landauer, T. K., & Harshman, R. (1990). Indexing by latent semantic analysis. *Journal of the American Society for Information Science, 41*(6), 391.

Deshpande, M., & Karypis, G. (2004). Item-based top-n recommendation algorithms. *ACM Transactions on Information Systems (TOIS), 22*(1), 143–177.

Doush, I. A., Alkhateeb, F., Maghayreh, E. A., Alsmadi, I., & Samarah, S. (2012). Annotations, collaborative tagging, and searching mathematics in e-learning. *arXiv Preprint.* arXiv:1211.1780.

Godoy, D., & Amandi, A. (2008). Hybrid content and tag-based profiles for recommendation in collaborative tagging systems. In *Proceedings of the Latin American Web Conference, LA-WEB 2008* (pp. 58–65). http://doi.org/10.1109/LA-WEB.2008.15

Golder, S. A., & Huberman, B. A. (2006). The structure of collaborative tagging systems. *Journal of Information Science, 32*(2), 198–208.

Gordon-Murnane, L. (2006). Social bookmarking, folksonomies, and web 2.0 tools. *Red Orbit.* http://www.Encyclopedia.com/doc/1G1-146693738.Html, March 2009).

Guy, M., & Tonkin, E. (2006). Tidying up tags. *D-Lib Magazine, 12*(1), 1082–9873.

Halpin, H., Robu, V., & Shepherd, H. (2007). The complex dynamics of collaborative tagging. In *Proceedings of the 16th International Conference on World Wide Web* (pp. 211–220).

Herlocker, J. L., Konstan, J. A., Terveen, L. G., & Riedl, J. T. (2004). Evaluating collaborative filtering recommender systems. *ACM Transactions on Information Systems.* http://doi.org/10.1145/963770.963772

Hofmann, T. (1999). Probabilistic latent semantic indexing. In *Proceedings of the 22nd Annual International ACM SIGIR Conference on Research and Development in Information Retrieval* (pp. 50–57).

Hofmann, T. (2003). Collaborative filtering via Gaussian probabilistic latent semantic analysis. In *Proceedings of the 26th Annual International ACM SIGIR Conference on Research and Development in Information Retrieval* (pp. 259–266).

Hotho, A., Jäschke, R., Schmitz, C., & Stumme, G. (2006a). Information retrieval in folksonomies: Search and ranking. In *Lecture Notes in Computer Science (Including Subseries Lecture Notes in Artificial Intelligence and Lecture Notes in Bioinformatics)* (Vol. 4011 LNCS, pp. 411–426). http://doi.org/10.1007/11762256_31

Hotho, A., Jäschke, R., Schmitz, C., & Stumme, G. (2006b). BibSonomy: A social bookmark and publication sharing system. In *Proceedings of the Conceptual Structures Tool Interoperability Workshop at the 14th International Conference on Conceptual Structures* (Vol. 87, p. 102).

Hotho, A., Jäschke, R., Schmitz, C., Stumme, G., & Althoff, K.-D. (2006c). Folkrank: A ranking algorithm for folksonomies. In *LWA* (Vol. 1, pp. 111–114).

Jäschke, R., Marinho, L., Hotho, A., Schmidt-Thieme, L., & Stumme, G. (2007). Tag recommendations in folksonomies. In *Knowledge discovery in databases: PKDD 2007 SE—52* (Vol. 4702, pp. 506–514). http://doi.org/10.1007/978-3-540-74976-9_52

Liang, H., Xu, Y., Li, Y., & Nayak, R. (2008). Collaborative filtering recommender systems using tag information. In *Proceedings—2008 IEEE/WIC/ACM International Conference on Web Intelligence and Intelligent Agent Technology—Workshops, WI-IAT Workshops 2008* (pp. 59–62). http://doi.org/10.1109/WIIAT.2008.97

Lin, X., Beaudoin, J. E., Bui, Y., & Desai, K. (2006). Exploring characteristics of social classification. *Advances in Classification Research Online, 17*(1), 1–19.

MacLaurin, M. B. (2010). Selection-based item tagging. Google Patents.

Maier, R., & Thalmann, S. (2008). Institutionalised collaborative tagging as an instrument for managing the maturing learning and knowledge resources. *International Journal of Technology Enhanced Learning, 1*(1–2), 70–84.

Manouselis, N., Drachsler, H., Vuorikari, R., Hummel, H., & Koper, R. (2011). Recommender systems in technology enhanced learning. In *Recommender systems handbook* (pp. 387–415). http://doi.org/10.1007/978-0-387-85820-3

Marinho, L. B., Nanopoulos, A., Schmidt-Thieme, L., Jäschke, R., Hotho, A., Stumme, G., et al. (2011). Social tagging recommender systems. In *Recommender systems handbook* (pp. 615–644). Berlin: Springer.

Marinho, L. B., & Schmidt-Thieme, L. (2008). Collaborative tag recommendations. In *Data analysis, machine learning and applications* (pp. 533–540). Berlin: Springer.

Marlow, C., Naaman, M., Boyd, D., & Davis, M. (2006). HT06, tagging paper, taxonomy, Flickr, academic article, to read. In *Proceedings of the Seventeenth Conference on Hypertext and Hypermedia* (pp. 31–40).

Mathes, A. (2004). Folksonomies—Cooperative classification and communication through shared metadata. *Computer Mediated Communication—LIS590CMC*, 1–13. http://doi.org/10.1.1.135.1000

Mika, P. (2005). Ontologies are us: A unified model of social networks and semantics. In *The Semantic Web–ISWC 2005* (pp. 522–536). Berlin: Springer.

Milicevic, A. K., Nanopoulos, A., & Ivanovic, M. (2010). Social tagging in recommender systems: A survey of the state-of-the-art and possible extensions. *Artificial Intelligence Review, 33*, 187–209. http://doi.org/10.1007/s10462-009-9153-2

Millen, D. R., Feinberg, J., & Kerr, B. (2006). Dogear: Social bookmarking in the enterprise. In *Proceedings of the SIGCHI Conference on Human Factors in Computing Systems* (pp. 111–120).

Noll, M. G., Au Yeung, C., Gibbins, N., Meinel, C., & Shadbolt, N. (2009). Telling experts from spammers: Expertise ranking in folksonomies. In *Proceedings of the 32nd International ACM SIGIR Conference on Research and Development in Information Retrieval* (pp. 612–619).

Page, L., Brin, S., Motwani, R., & Winograd, T. (1999). The PageRank citation ranking: Bringing order to the web.

Palmatier, R. A., & Bennett, J. M. (1974). Notetaking habits of college students. *Journal of Reading, 18*(3), 215–218.

Peters, I., & Stock, W. G. (2007). Folksonomy and information retrieval. *Proceedings of the American Society for Information Science and Technology, 44*(1), 1–28.

Pierrakos, D., Paliouras, G., Papatheodorou, C., & Spyropoulos, C. D. (2003). Web usage mining as a tool for personalization: A survey. *User Modeling and User-Adapted Interaction, 13*(4), 311–372.

Pluzhenskaia, M. (2006). Folksonomies or fauxsonomies: How social is social bookmarking. In *17th ASIS&T SIG/CR Classification Research Workshop. Abstracts of Posters (S. 23-24).*

Quintarelli, E. (2005). Folksonomies: power to the people.

Rendle, S., Balby Marinho, L., Nanopoulos, A., & Schmidt-Thieme, L. (2009). Learning optimal ranking with tensor factorization for tag recommendation. In *Proceedings of the 15th ACM SIGKDD International Conference on Knowledge Discovery and Data Mining* (pp. 727–736).

Resnick, P., Iacovou, N., Suchak, M., Bergstrom, P., & Riedl, J. (1994). GroupLens: An open architecture for collaborative filtering of netnews. In *Proceedings of the 1994 ACM Conference on Computer Supported Cooperative Work* (pp. 175–186).

Santos, O. C., & Boticario, J. G. (2008). Intelligent support for inclusive eLearning. In *Proceedings—2008 IEEE/WIC/ACM International Conference on Web Intelligence and Intelligent Agent Technology—Workshops, WI-IAT Workshops 2008* (pp. 361–364). http://doi.org/10.1109/WIIAT.2008.372

Shardanand, U., & Maes, P. (1995). Social information filtering: Algorithms for automating "Word of Mouth." In *ACM Conference on Human Factors in Computing Systems (CHI)* (Vol. 1, pp. 210–217). http://doi.org/10.1145/223904.223931

Shepitsen, A., Gemmell, J., Mobasher, B., & Burke, R. (2008). Personalized recommendation in social tagging systems using hierarchical clustering. In *Proceedings of the 2008 ACM Conference on Recommender systems* (pp. 259–266).

Sood, S. C., Owsley, S. H., Hammond, K. J., & Birnbaum, L. (2007). TagAssist: Automatic tag suggestion for blog posts. In *Proceedings of the 1st International Conference on Weblogs and Social Media (ICWSM 2007)* (pp. 1–8).

Symeonidis, P., Ruxanda, M. M., Nanopoulos, A., & Manolopoulos, Y. (2008). Ternary semantic analysis of social tags for personalized music recommendation. In *ISMIR* (Vol. 8, pp. 219–224).

Tan, H., Guo, J., & Li, Y. (2008). E-learning recommendation system. In *2008 International Conference on Computer Science and Software Engineering,* (Vol. 5, pp. 430–433).

Tatu, M., Srikanth, M., & D'Silva, T. (2008). Rsdc'08: Tag recommendations using bookmark content. *ECML PKDD Discovery Challenge, 2008*, 96–107.

Tso-sutter, K. H. L., Marinho, L. B., & Schmidt-Thieme, L. (2008). Tag-aware recommender systems by fusion of collaborative filtering algorithms. *Search*, 1995–1999. http://doi.org/10.1145/1363686.1364171

Veres, C. (2006a). Concept modeling by the masses: Folksonomy structure and interoperability. In *Conceptual Modeling-ER 2006* (pp. 325–338). Berlin: Springer.

Veres, C. (2006b). The language of folksonomies: What tags reveal about user classification. In *Natural language processing and information systems* (pp. 58–69). Berlin: Springer.

Wal, T. Vander. (2007). Folksonomy coinage and definition. *Vanderwalnet.*

Wetzker, R., Said, A., & Zimmermann, C. (2009). Understanding the user: Personomy translation for tag recommendation. *ECML PKDD Discovery Challenge, 2009.*

Wiley, D. (2007). Connecting learning objects to instructional design theory: A definition, a metaphor, and a taxonomy. The Instructional Use of Learning Objects: Online Version. 2000. *Available on Web Site*: http://reusability.Org/read/chapters/wiley.Doc

Winget, M. (2006). User-defined classification on the online photo sharing site Flickr... Or, how I learned to stop worrying and love the million typing monkeys. *Advances in Classification Research Online, 17*(1), 1–16.

Wu, S., Ghenniwa, H., Zhang, Y., & Shen, W. (2006). Personal assistant agents for collaborative design environments. *Computers in Industry, 57*(8–9), 732–739. http://doi.org/10.1016/j.compind.2006.04.010

Xu, Z., Fu, Y., Mao, J., & Su, D. (2006). Towards the semantic web: Collaborative tag suggestions. In *Collaborative Web Tagging Workshop at WWW2006, Edinburgh, Scotland.*

Semantic Web Technologies in E-Learning

Part III
Semantic Web Technologies in E-Learning

Chapter 8
Semantic Web

Abstract The Semantic Web is a next generation of the Web in which information is presented in such a way that it can be used by computers not only to be presented but also to be used for automation of the search, integration, and reuse between applications. The goal of the Semantic Web is to develop the basis for intelligent applications that enable more efficient information use by not just providing a set of linked documents but a collection of knowledge repositories with meaningful content and additional logic structure. Also, one of the main goals is to build an appropriate infrastructure for intelligent agents to perform complex actions on the network. There are a number of important concepts that enable the development of the Semantic Web. This chapter presents the most important of them: knowledge organization systems, ontologies, Semantic Web languages and adaptation rules. Possibilities of applying Semantic Web technologies in e-learning systems are presented in this chapter.

The Semantic Web has emerged as a vision of Tim Berners-Lee,[1] the global Web as a universal medium for data, information and knowledge (Berners-Lee 2000). In 1999, he expressed his vision of the Semantic Web as follows: *"I have a dream for the Web (in which computers) become capable of analysing all the data on the Web —the content, links, and transactions between people and computers. A Semantic Web, which makes this possible, has yet to emerge, but when it does, the day-to-day mechanisms of trade, bureaucracy and our daily lives will be handled by machines talking to machines. The intelligent agents, people have touted for ages will finally materialize"*.

People in their daily activities use the Web to do various tasks such as finding the meaning of foreign words, the provision of books in the library, search the lower prices of certain items, and so on. However, the computer cannot fulfil these tasks without human guidance because the Web pages are designed to be read by people, not computers. Therefore, the Semantic Web can be considered as an attempt to display and store information that are understandable to computers so that they can

---

[1]Berners-Lee is the director of the World Wide Web Consortium (W3C). The World Wide Web Consortium (W3C) is the main international standards organization for the World Wide Web.

© Springer International Publishing Switzerland 2017   115
A. Klašnja-Milićević et al., *E-Learning Systems*,
Intelligent Systems Reference Library 112, DOI 10.1007/978-3-319-41163-7_8

work with demanding tasks involved in finding, sharing and exchanging information online.

The Semantic Web is a next generation of the Web in which information is presented in such a way that it can be used by computers not only for display but also to automate the search, integration, and reuse between applications (Alsultanny 2006). The goal of the Semantic Web is to develop the basis for intelligent applications that enable more efficient information use by not just providing a set of linked documents but a collection of knowledge repositories with meaningful content and additional logic structure (Sheth et al. 2005). Also, one of the main goals is to build an appropriate infrastructure for intelligent agents to perform complex actions on the network. In order to do that, agents must retrieve and process the relevant information. That process requires unconditional integration of agents and networks, and full use of the advantages offered by the existing infrastructure. Further, the Semantic Web is an explicitly declared and integrated into Web applications that provides semantic access and extract of relevant information from the text. Finally, the Semantic Web is a way to reliably implement deep integration of Web services that are understandable to computers. This creates a network of tightly coupled services, that will intelligent agents be able to discover, execute and automatically combine (Berners-Lee et al. 2001).

The problem with today's Web is that it is huge, but not enough "smart" to easily integrate information, needed by the user. Such integration is required in almost all forms of Web use. Most of the data from the Web is represented in natural language and as such it is not understandable to computers. On the other hand, people can only handle a small portion of the information from the Web and it would be useful if computers would take over part of the job for processing and analysing content from the Web. Unfortunately, the Web is generally oriented and is intended for human use, which means that all content on the Web is readable for computers but not understandable. Users need Semantic Web to display information in an accurate, computer understandable way, in a form suitable for sharing, reuse and processing by software agents. Explicit representation of the metadata enables the development of the Web that offers completely new forms of services, such as, for example, the intelligent search techniques and efficient exchange and filtering information.

There are a number of important concepts that enable the development of the Semantic Web. The following sections will present the most important concepts like: knowledge organization systems, ontologies, Semantic Web languages and adaptation rules.

8.1 Knowledge Organization Systems

Independently of an area or domain of human being activities and efforts like: meteorology or bank transactions, proteins or engine parts, concepts are needed (Shadbolt et al. 2006). Collected data about definitions, systematisations and structuring of used concepts in different domains must be structured, organized and allow easy access. The term knowledge organization systems is intended to incorporate all types of structures for organizing information and promoting knowledge management (Hodge 2000). Knowledge organization systems are used to organize materials for the purpose of retrieval and easier managing. These systems serve as a bridge between the user's need for information and the available information and knowledge.

Modern technologies have enabled the appearance of the new generation of knowledge systems and they should be able to quickly and efficiently complete their tasks (Gruber 2008):

- **Capture new information**. Accessible and cheap sensors, microprocessors, memory, fiber networks, and cellular telephony influenced that most of users have computers, smart mobile phones, digital cameras, and broadband Internet access. These things enable users to upload their digital lives and spend more time online.
- **Store acquired information**. Cheap disk storage allows sharing huge amounts of information among people.
- **Distribution of information**. The Internet is an information superconductor connecting the planet.
- **Communication**. Asynchronous collaboration systems (email, wikis and blogs) overcome barriers of space and time and other limitations for conversation. People can easily communicate and share knowledge.

Initially, artificial intelligence techniques were applied for knowledge reengineering to achieve the semantics, but maintaining the knowledge bases is not an easy task either (Babu and Krishnamurthy 2013). Even more, the manifestation of digital libraries as one of the important knowledge based system has thrown up potential challenges for knowledge acquisition as well as integrating with wide range of intelligent applications (d'Aquin et al. 2008).

Various structures and features of a knowledge organization system are used for building online repositories. New varieties of knowledge organization systems, like, taxonomies, vocabularies and ontologies were brought into existence, mostly to serve specific functions (Gilchrist 2000).

8.2 Ontologies

The word *ontology* is of Greek origin and is derived from the word *ontos* which means being or existing and *logos* meaning science or learning. In philosophy it presents the unit of existence and study of the properties of elements. More specifically, the word ontology is the study of the things that exist in a particular area or domain (Gruber 1995).

Informally, the ontology of a certain domain regards terminology (domain vocabulary), all important concepts in the domain, their classification, taxonomy, their connections (including the all relevant hierarchy and constraints) and all the axioms of the domain (Devedžić 2004). Formally, for someone who wants to discuss matters of a domain D using a language L, ontology provides catalogue of elements that exist in a domain D. These types of ontology have been presented in the form of concepts, relations and predicates defined in the language L. Either formally or informally, ontology is a very important part of any knowledge domain. Ontology is often a key part of the knowledge and all the other knowledge must rely and refer to it.

Ontologies are a glossary of terms whose semantics are formally defined. They formally and explicitly define concepts within a particular domain, the relationships between these concepts and their properties (Gascueña et al. 2006). Also, the ontologies are useful in various domains since they allow people and software agents to understand presented information and structure of the knowledge. Reuse of ontologies and its individual parts are also enabled, i.e. development of a new ontology is not required if appropriate ontologies are already present in the area.

In artificial intelligence, the term ontology is used for one of two related things (Chandrasekaran et al. 1999):

- presentation language, often specialized for a particular domain or topic,
- segment of knowledge that describes a domain with appropriate presentation language.

In both cases, there is always a corresponding data structure that represents the ontology.

Semantic Web technologies seem to be a promising technological foundation for the next generation of e-learning systems (Breslin et al. 2011). Ontology, generally defined as a representation of a shared conceptualization of a particular domain, is one of essential components of the Semantic Web. The initial work on implementing ontologies as the backbone of e-learning systems is presented in (Mizoguchi and Bourdeau 2015). Since that time, many authors have proposed the usage of ontologies for different purposes in e-learning environments, such as adaptive hypermedia, personalization, and learner modelling (Jovanovic et al. 2007).

Ontologies provide a vocabulary of terms whose semantics are formally specified. Interest in ontologies has also grown as researchers and system developers have become more interested in reusing and sharing knowledge across wide range of systems (Swartout & Tate, 1999). Currently, one key obstacle to sharing knowledge is that different systems use different concepts and terms for describing domains. These differences make it difficult to take knowledge out of one system and use it smoothly in another. If we could develop ontologies that might be used as the basis for multiple systems, they would share a common terminology that would facilitate sharing and reuse of knowledge. Developing such reusable ontologies is an important goal of ontology research.

8.2.1 Adaptive Educational Systems Technologies in E-Learning

Building adaptive educational system requires a lot of effort and often is done from scratch. It becomes even more demanding with the constant increase of the information available on the Web and with the involvement of complex adaptation strategies for the instructional content presentation and navigation (Aroyo and Mizoguchi 2003).

Important concern about the Semantic Web is the cost of ontology development and maintenance (Shadbolt et al. 2006). In some areas, the costs would be easy to justify. For example, an ontology could be a powerful and essential tool in well-structured areas such as scientific applications.

In order to provide a richer set of educational functionalities and increase effectiveness of distance learning, educational systems need to interoperate, collaborate and exchange content or re-use functionality (Aroyo and Dicheva 2004). Key tools for enabling the interoperability are:

- semantic conceptualization and ontologies,
- common standardized communication syntax, and
- large-scale service-based integration of educational content and functionality provision and usage.

In certain commercial applications, the potential benefit and productivity gain from using well-structured and coordinated vocabulary specifications will compensate the costs of ontology development and the costs of maintenance. In fact, costs might decrease as an ontology's user base increases, and number of required ontology engineers increases as the user community grows. The consequence is that the effort involved per user in building ontologies for large communities becomes very small very quickly (Shadbolt et al. 2006).

8.2.2 Standards for E-Learning Environments

The Semantic Web offers new technologies to the developers of e-learning systems in order to provide more intelligent access and management of learning material, as well as semantically richer modelling of the learners (Aroyo and Dicheva 2004). The learning technology community is quickly adopted to many of the Semantic Web technologies. Determining of how researchers and practitioners are using semantic technologies are crucial for apprehension of certain key trends in the Semantic Web (Cardoso 2007).

An ultimate goal of e-learning system developers nowadays is to create information and knowledge components that are easily accessible and usable by numerous learners (Aroyo and Dicheva 2004). For example, many researches demand the integration of diverse and heterogeneous data sets that originate from distinct communities of scientists in separate subfields (Shadbolt et al. 2006). In order to integrate vast amount of data from various sources, these data should follow previously agreed formats and standards. Scientists, researchers, and regulatory authorities from various areas need a way to integrate these data (Shadbolt et al. 2006). This is being achieved in large part through the adoption of common conceptualizations in the form of ontologies.

As need for implementation of Web semantics grows, the user community, including organization like the World Wide Web Consortium (W3C), has directed major efforts at specifying, developing, and deploying languages for sharing information (Shadbolt et al. 2006). The ultimate goal for these languages was to provide a foundation for semantic interoperability. The W3C defined the first Resource Description Framework (RDF) specification in 1997. RDF provided a simple but powerful representation language for *Universal Resource Identifiers* and became an official W3C recommendation. That was a crucial step in drawing attention to the specification and promoting its widespread deployment to enhance the Web's functionality and interoperability (Shadbolt et al. 2006).

There is a growing concern towards the need of extending the existing educational standards, such as the IEEE/IMS LOM standard, in the context of the Semantic Web so as to allow improved semantic annotation of learning resources (Aroyo and Dicheva 2004). Significant evolution of standards as improvements and innovations will allow the delivery of more complex and sophisticated semantic applications (Cardoso 2007). Numerous initiatives are oriented towards developing ontologies for biology, medicine, distance learning, and other related fields. These communities have been developing language standards that can be deployed on the Web (Shadbolt et al. 2006).

8.2.3 Semantic Web Methodologies

The common problems in educational environments are related to keeping up with the constant requirements for flexibility and adaptability of content and for reusability and sharing of learning objects and structures (Devedzic and Harrer 2005). Another problem in the current web-based educational systems research is that assessment of the existing systems is difficult as there is no common reference architecture, nor standardized approaches. Thus, there is an increasing need for efficient support for the designers and authors of adaptive educational systems (Aroyo and Dicheva 2004).

Until a few years ago the building of ontologies was done in a rather ad hoc fashion. Meanwhile, there have been some few, but influential proposals for guiding the ontology development process (Staab et al. 2001). In contrast to other methodologies (Guarino and Welty 2000), which mostly restrict their attention within the ontology itself, authors in (Staab et al. 2001) presented approach that focuses on the application-driven development of ontologies. Their methodology cover majority of aspects: starting from the early stages of setting up a knowledge management project to the final roll out of the ontology-based application. This methodology is comprised of following phases (Staab et al. 2001):

- **Feasibility study**. Knowledge management system is efficient only if it is properly integrated into the organization in which it is operational. Many factors other than technology determine success of such a system. To analyse these factors, feasibility study must be performed in order to identify problem/opportunity areas and potential solutions, and later, to put them into a wider organizational perspective. The feasibility study serves as a decision support for economic and technical project feasibility, in order to select the most promising focus area and target solution (Staab et al. 2001).
- **Kickoff phase for ontology development**. The result of this phase is an ontology requirements specification document that describes what an ontology should support and it describes ontology application. It should also guide an ontology engineer to decide about inclusion, exclusion and the hierarchical structure of concepts in the ontology. In this early stage, one should look for already developed and potentially reusable ontologies (Staab et al. 2001).
- **Refinement phase**. The goal of this phase is to produce a mature and application-oriented target ontology according to the specification given in the kickoff phase (Staab et al. 2001).
- **Evaluation phase**. The evaluation phase serves as a proof for the usefulness of developed ontologies and their associated software environments. In a first step, the ontology engineer checks, whether the created ontology fulfils defined ontology requirements specification and whether the ontology supports the competency questions analysed in the kick-off phase of the project. In a second step, the ontology is tested in the target application environment. Later, feedback from users may be a valuable input for further refinement of the ontology (Staab et al. 2001).
- **Maintenance phase**. In the real world and environments, different aspects and things are constantly changing—and so does specifications for ontologies. To reflect these changes ontologies have to be maintained frequently like other parts of software, too. The maintenance of ontologies is primarily an organizational process that must have precisely defined rules for the update processes within ontologies (Staab et al. 2001).

8.2.4 Representation of Ontologies

People describe ontologies as sets of declarative statements in natural language. However, the statements in a natural language are difficult for computers to process. Ontology presentation in computers requires formal languages. Ontologies have various forms depending on the level of abstraction. When implemented in a computer, they usually appear as XML files (Bray et al. 1998). Since ontologies always represent some concepts and relations between them, they can be graphically depicted by visual languages. Graphical tools for building ontologies always support the conversion from graphic formats to XML and other textual representation of the data.

Ontologies are the basis for sharing conceptualization of a domain, semantic annotation and design of concepts, their relationships and properties (Sivashanmugam et al. 2003). Ontologies aim at capturing domain knowledge in a generic way and provide a commonly agreed understanding of a domain. This knowledge may be reused and shared across applications (Staab et al. 2001). Ontologies as form of knowledge organization in systems are used for:

- data retrieval and directing an user toward documents of his/her interest more efficiently,
- data browsing,
- navigation through documents, locating and identification of information,
- providing the user with general overview of knowledge structure in domain,
- providing arranged sequences or orders of existing documents.

Many technologies have been introduced for the implementation of Semantic Web that allows creating, storing and linking data, building vocabularies, and defining rules for handling this data (Horrocks 2008). Some of the standardized technologies that allows linked data process are:

- RDF—Resource Definition Framework
- OWL—Web Ontology Language
- SKOS—Simple Knowledge Organization System
- RIF—Rule Interchange Format
- etc.

It is not a question which language is best for building elements of the Semantic Web, the goal is to find the most suitable language for the representation of Semantic Web elements (Gómez-Pérez and Corcho 2002). Not all of these languages are useful for different applications: each of them will require different specifications since, for instance, they have different complexity levels. For example, authors in (Cardoso 2007) presented that the Semantic Web does not require complex ontologies and that large majority of developed ontologies are rather small. They showed that the Semantic Web does not even need OWL and can achieve important objectives such as data-sharing and data-integration using just RDF alone.

8.2.4.1   Example: Ontology of an E-Learning System

Let's give an example of learner modelling in a simple e-learning system. System contains some basic concepts such as *learner, course, lesson, course author* and *mentor*. Only certain relations between concepts are presented: the author of course creates a course, mentor prepares lessons, learner attends lessons that are part of the course, etc. This is a series of declarative sentences that represent ontology concepts and their relationships. These sentences are only understandable to people. As such, they are not formally defined and adapted for computer processing.

At a higher level of abstraction, this ontology with its concepts can be informally displayed as a semantic network diagram (Fig. 8.1).

Previous diagram suffers from many drawbacks, such as its informality, lack of detail, etc. To make this ontology presented more formally, UML diagram can be used (Fig. 8.2). This diagram represents the same part of the domain but formally presented and in more detail.

Figure 8.3 shows the same ontology in XML-based format with use of Web Ontology Language (OWL) language, which will be discussed in more detail in this chapter. Ontology graphic editor automatically generates the appropriate part of the code. This representation of ontologies is most commonly used at the implementation level of ontologies in systems.

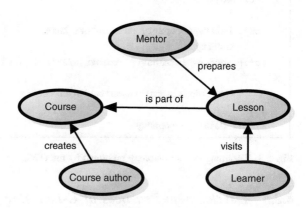

Fig. 8.1 E-learning system ontology presented as semantic network

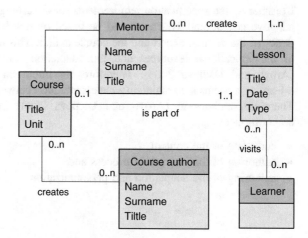

Fig. 8.2 UML class diagram of e-learning system

```
<owl:Class rdf:ID="Course"/>
<owl:Class rdf:ID="Author"/>
<owl:Class rdf:ID="Mentor"/>
<owl:Class rdf:ID="Learner"/>
<owl:Class rdf:ID="Lesson"/>
<owl:ObjectProperty rdf:ID="creates">
      <owl:inverseOf>
              <owl;ObjectProperty rdf;ID="createdBy"/>
      </owl:inverseOf>
      <rdfs:domain rdf:resource="#Author"/>
      <rdfs:range rdf:resource="#Course"/>
</owl:ObjectProperty>
<owl:ObjectProperty rdf:ID="Prepares">
      <rdfs:range rdf:resource="#Lesson"/>
      <rdfs;domain rdf:resource="#Mentor"/>
</owl:ObjectProperty>
...
<owl: DatatypeProperty rdf:ID="LastName">
      <rdf:type
rdf:resource="localhost/ontologija/2012/0"//owl#FunctionalProperty"/
>
      <rdfs:domain rdf:resource="#Learner"/>
...
</owl: DatatypeProperty>
```

Fig. 8.3 E-learning system ontology presented by the OWL

8.2.5   Development Practices of E-Learning Systems

Considering the specification requirements for e-learning systems, the authoring of adaptive educational system should be based on a strict separation and independency of the domain expert and the course author. This implies separate definitions of the educational resources and the adaptation and personalization strategies (Aroyo and Dicheva 2004). Therefore, an important goal for engineering of e-learning systems is to maintain a clear separation between concepts and resources. Therefore, authors in (Aroyo and Dicheva 2004) identified three groups of authoring activities:

- authoring of the content,
- authoring of the instruction process and
- authoring of the adaptation and personalization.

8.2.5.1   Authoring of Content

Educational ontologies typically consist of concepts definitions, relations and axioms (Staab et al. 2001). Authoring of educational content in e-learning systems concerns creation of learning objects, their annotation, and links between them. Content authors in distance learning systems can explicitly define the semantics of resources within domain. At this level, the authors perform following activities (Aroyo and Dicheva 2004):

- **Domain-related authoring activities**. These activities include constructing and editing of the domain model that includes creation of concepts and their links. Concept from a domain is defined as a pair, consisting of the concept's name and the corresponding set of attributes. A link presents an association between two concepts of a certain type with a specific weight and specified link direction.
- **Resource-related authoring activities**. These activities include building a collection of educational resources and creating of a resource repository. Each resource is enriched with the appropriate metadata to facilitate its further use within the course sequencing module.

Therefore, Semantic Web applications consist of two separate but linked layers (Knublauch 2004):

- **Semantic Web Layer** that makes ontologies and interfaces available to the public, and
- **Internal Layer** which consists of the control and reasoning mechanisms.

While the latter components can reside inside a *black box*, the artefacts in the Semantic Web Layer are shared with other applications, and must therefore meet higher quality standards than the internal components (Knublauch 2004).

8.2.5.2   Authoring of Instructional Process

Authoring of the instructional process in educational systems typically involves course construction activities (generating a course tasks model) that serve as a basis for the further sequencing of course tasks (Aroyo and Dicheva 2004). In order to produce an instructional task sequence the author usually:

- selects concepts from the domain model and assigns them to course topics
- selects specific sequences of course topics realizing the learning goals, and
- assigns course tasks for each topic (each task will cover more than one learning activity).

8.2.5.3   Authoring of the Adaptation and Personalization

These authoring activities ensure structuring and editing of domain concepts and resources, modelling of the course process, task sequencing and selecting and applying an appropriate adaptation strategy (Aroyo and Dicheva 2004). With the new frameworks and architectures that are evolving in order to meet the semantic challenge, the goal has become to provide the learners and course authors with a seamless personalized interaction with the educational systems (Aroyo and Dicheva 2004).

8.2.6   The Objective of Ontologies

Regardless of the fact that ontologies have been used for many years in various domains and fields, there is still an open question: "Why ontologies?" One answer might be: because they are the basic building blocks of the systems and applications that supports the Semantic Web. Semantic cooperation between Web applications is possible only if the semantics of Web data are explicitly displayed on the Web in the form of a computer understandable and theoretical content domain—ontology. With the automated use and computer interpretation of ontology, computers themselves may offer unlimited support and automate the access and processing of information on the Web. Therefore, use of ontologies takes burden off the user and most of the work is carried out by the Semantic Web (Fensel and Musen 2001).

Ontologies provide access to global collection of human knowledge that computers can understand and process. Once the knowledge from a certain domain is put on a Web in the form of related ontologies, it creates the basis for further development of intelligent applications in that particular field, because it facilitates the problem of knowledge acquisition.

Ontologies play multiple roles in architecture of the Semantic Web (Horrocks et al. 2005):

- Ontologies allow processing, sharing, and reusing knowledge from the Web. This is achieved by using mutual concepts and terminology between different applications.
- They provide a higher level of interoperability on the Web, as they provide various forms of mapping data, because these mappings require semantic analysis.
- They allow the use of intelligent services—agents for search, information filtering, intelligent information integration, knowledge management, etc.

Ontologies are used to build many useful elements of intelligent systems, both in the general representation and in the process of knowledge engineering. Some of these basic elements are (Gruber 1995):

- *Vocabulary*. Ontologies provide a vocabulary of terms in the relevant field. These vocabularies are different from natural vocabularies (where words can be interpreted differently depending on the people who interpret them) as they offer logical statements that describe the terms and their relations. Also, ontologies define rules for combining these terms and their relationships in order to expand vocabulary. Ontologies define terms in a unique and unambiguous way, independent on the environment or the person who is observing them.
- *Concept hierarchy*. Taxonomy or concepts hierarchy is hierarchical categorization or classification of the entities within the relevant domain. Each ontology presents taxonomy in a computer-understandable way. Specific relations among concepts are formally defined in ontologies and provide consistency in the use of ontologies.
- *Content theory*. Since ontologies identify classes of objects, their relationships and hierarchies that exist in a domain, they are, in fact, authentic *Content theory*. Ontologies not only identify classes, relationships and hierarchies, but they are descriptive and determined by use of appropriate *Ontology representation languages*. These languages will be discussed later in this chapter.
- *Consistency checking*. Well-structured and processed ontologies allow different checks of correctness and consistency (type and value checking). They also allow portability between different applications that can perform this verification.
- *Reuse and sharing of knowledge*. The most important role of ontologies is not display of vocabularies and hierarchies, but enabling reuse and sharing of knowledge between different applications. The aim is to offer a description of each concept and relationship that exist within the domain and make them available to a variety of intelligent agents and applications.

Although there are many languages for presenting ontologies and tools for knowledge base development, easy sharing and reuse of ontologies is not easy to achieve as there are different ontologies that present the same knowledge but are written in different languages.

8.2.7 Ontology Application

Although it is hard to predict future trends in e-learning, it is obvious that the current vision of the Educational Semantic Web provides interoperability and reusability, based on implemented semantics, standardized communication among modular and service-oriented systems (Aroyo and Dicheva 2004). Course authors have few tools to easily and efficiently generate Semantic Web annotations using OWL, RDF or other languages concerning content use or creation (Shadbolt et al. 2006). OWL is most widely used but still needs additional tools and software development environments to support its production and application. An essential element for success is the availability of support for user-friendly, structured and automated authoring of educational systems, with balance of exploiting explicit semantic information and collecting and maintaining its semantics (Aroyo and Dicheva 2004). Modern ontology development tools such as Protégé (with the OWL Plugin) allow users to design and implement Semantic Web components and provide intelligent tools for finding mistakes, similar to a debugger in a programming environments (Knublauch 2004).

Some practices for development of software for the Semantic Web are presented in (Knublauch 2004). Authors demonstrated a realistic example scenario from the tourism domain, presented software architecture for those applications and suggested appropriate development guidelines. Dealing with ontologies and concepts increases our conceptual awareness and influences the style of information perception, which reflects in the demands for using and authoring web based educational systems (Aroyo and Dicheva 2004).

However, not only are ontologies useful for applications in which knowledge plays a key role, but rules and inference are also implemented and used (Gómez-Pérez and Corcho 2002). The OWL language itself is designed to support various types of inference—typically, inheritance and classification. Because it's difficult to specify a formalism that will capture all the knowledge in a particular domain, there are other approaches to inference on the Web. Work has begun on the *Rule Interchange Format*, an attempt to support and interoperate across a variety of rule-based formats (Shadbolt et al. 2006).

Numerous areas of ontologies application can be classified into four categories: cooperation, interoperability, education and modelling (Sheth et al. 2005).

Cooperation. Different people may have different views on the same problem when working on a team project. For them, ontologies offer a unique skeleton of knowledge as a guideline for further development. Even more important role of ontologies is enabling cooperation and communication between intelligent agents. Knowledge exchange between different agents is clearer when agents are aware of ontologies that other agents are using as data models.

Interoperability. Ontologies allow the integration of information from different and separate sources. Users typically do not care how it will get the information; the more important for them is to get all the information they need. Different applications are often required to access various sources of knowledge in order to get to all the available information. These different sources may contain information in different formats and at different level of detail. Therefore, if all sources identify the same ontology, then data conversion and integration of information can be easily performed and automated.

Education. Ontologies are also a good medium for the publication of scientific papers and creation of references database. Since they usually occur on the basis of consensus about the knowledge of the domain, ontologies can provide reliable and objective information for those who want to learn more about a given domain.

Modelling. Ontologies are important re-usable building blocks for intelligent applications modelling, and can be includes as pre-existing modules of knowledge.

There are some other categories of ontology use such as in electronic commerce (to enable communication between sellers and buyers) and for automated search.

8.3   Semantic Web Languages

There are common attempts to standardize the form of information in the various components of the e-learning systems (Aroyo and Mizoguchi 2003). However, the differences are still present, since there are different proposals for standards and terminology for describing resources and activities.

In the literature, the terms *Web-based ontology languages* and *Semantic Web languages* are used interchangeably. Semantic Web languages are formal languages for creating ontologies and the Semantic Web. Semantic Web technology is built in layers, i.e. it is carried out in steps (Berners-Lee et al. 2001; Dutta 2006). Higher-level Semantic Web languages use the syntax and semantics of lower-levels languages (Fig. 8.4).

XML allows users to add arbitrary structure to their documents and to create tags for labelling Web pages or parts of the text (Bray et al. 1998). Although the meaning of the XML tags is intuitively clear, it does not allow adding semantics. XML is only used as a transport mechanism. *Resource Description Framework—* RDF and *Resource description framework schema—*RDFS provide a basic framework for displaying metadata on the Web. Other languages for knowledge representation on the Web, such as OWL (based on RDF), provide a more precise support for knowledge representation.

8.3.1   XML—eXtensible Markup Language

Important step towards implementation of syntax and semantics of data in applications is agreement upon their unique presentation. All Semantic Web languages use XML syntax in order to prevent unnecessary and costly modification of applications for bridging differences in semantics and syntax of the data. In fact,

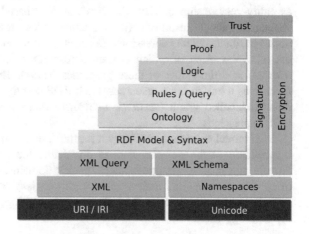

Fig. 8.4 Semantic Web layers (Berners-Lee et al. 2001; Dutta 2006)

XML is a meta-language for representation of Semantic Web languages (Bray et al. 1998). For example, XML Schema defines a class of XML documents using XML syntax. RDF provides a framework for representing metadata about Web resources, and can be represented in XML as well. RDF Schema, OWL and other Web ontology languages also use XML syntax.

Extensible Markup Language (XML) is a markup language that defines a set of rules for encoding documents in a format that is both human-readable and machine-readable. Each XML document contains one or more elements, which boundaries are delimited by start-tags and end-tags. Each XML Schema provides the necessary framework for creating categories of XML documents.

XML is a specification for computer-readable documents (Klein 2001). Markup means that certain sequences of characters in the document contain information indicating the role of the document's content.

XML does not indicate a specific computer data interpretation. The information in XML documents are only presented in unique syntax, but their use and semantics are not specified. In other words, XML defines only the structure of the documents, not their computational interpretation. It offers a structured document format, without specifying the dictionary. To represent knowledge and semantics more precise XML-based languages must be used.

8.3.2   RDFS—Resource Description Framework Schema

RDFS or RDF Schema is a language for the presentation of knowledge that provides basic elements for description of ontologies (also called RDF vocabularies) for the purpose of structuring resources (Brickley and Guha 2008). The first version of the language was published by W3C in April 1998 and the final recommendations were published in February 2004. The most important RDFS components are included in the expressive language OWL.

The RDF data model is similar to classic conceptual modelling approaches such as entity–relationship or class diagrams, as it is based upon the idea of making statements about resources (in particular Web resources) in the form of object-attributes-value expressions. Description of resources in RDF is presented as list of triples, each of which contain a resource (*object*), its characteristics (*attributes*), and the values of these properties (*value*). The value can be in the text form or as a link to other resources. Each RDF description can be represented as a directed labelled graph (*semantic network*), whose parts are equivalent to other RDF expressions.

RDF model provides a mechanism for the description of the individual resources independent of the domain. It does not define the semantics of an application domain, nor make assumptions about specific domains. In order to define elements of specific domains and their semantics (ontologies), other environments are needed. RDF is used to describe the instance of the ontology, and RDF Schema defines the ontology itself.

RDF Schema (or RDFS) is a set of classes with certain properties using the RDF extensible knowledge representation language, providing basic elements for the description of ontologies, otherwise called RDF vocabularies, intended to structure RDF resources.

RDF and RDFS allow the description of the facts about Web resources, but they often require richer and more accurate elements to define the formal semantics of Web resources. For example, in the RDFS, classes cannot be compared and it is impossible to define the constraints of cardinalities.

8.3.3   OWL—Ontology Web Language

The Web Ontology Language (OWL) is a family of knowledge representation languages or ontology languages for authoring ontologies or knowledge bases, endorsed by the *World Wide Web Consortium* (Bechhofer 2009; Brickley and Guha 2008). This family of languages is based on two (largely, but not entirely compatible) semantics: OWL DL and OWL Lite.

OWL Dictionary includes a set of XML elements and attributes, with a precisely defined meaning. They are used to describe the terms of domains and their relations in the ontology. OWL vocabulary is based on the RDF vocabulary. OWL also shares all the elements of the domain types (values belonging to XML data types) and domain objects (individual buildings like the appearance of classes defined in OWL or RDF). There are two kinds of OWL properties:

- properties that connect objects to other objects (defined with owl: ObjectProperty)
- properties that connect objects with values of data types (defined with owl: DatatypeProperty)

OWL allows the definition of additional constraints and relationships between resources, such as cardinality, domain restriction, union, intersection, inverse and transitive rules.

SPARQL—Simple Protocol and RDF Query Language

Unlike OWL and RDF, SPARQL is not intended for the representation of ontologies and resources, but for the selecting data on the Web. More specifically, the SPARQL is query language for RDF.

SPARQL can be used for:

- selecting information from RDF graphs,
- selecting RDF subgraphs and
- construct new RDF graphs based on information selected from existing RDF graphs.

SPARQL queries have syntax similar to traditional database query languages like SQL. Therefore, these queries have the keywords SELECT and WHERE.

8.4   Graphical Environments for Ontology Development

Graphical environments for ontology development integrate ontology editor with
other tools, and usually support use of multiple languages for ontology presentation. Their aim is to offer support for the whole ontology development process and
their subsequent use.

8.4.1   Protégé

Currently, the leading editor for ontology development is Protégé, developed at
Stanford University (Protégé 2011). Protégé is a graphical tool for ontology editing
and knowledge acquisition that can be adapted to enable conceptual modelling with
new and evolving Semantic Web languages. Protégé allows the definition of
ontology concepts (classes), their attributes, hierarchies and different constraints as
well as instances of classes (Fig. 8.5).

The Protégé OWL Plugin provides a SWRL editor, which enables the formalization of SWRL rules in conjunction with OWL ontologies. It provides graphical
user interface for easy development and management of ontologies. Moreover, the

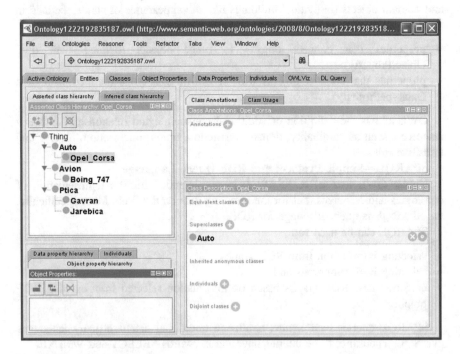

Fig. 8.5 Protégé ontology editor

SWRLTab within it provides editor and validation tool to develop inference rules. In addition, other ontologies can be imported to achieve knowledge reuse.

Protégé can be used for creation of domain models at a conceptual level without having to know the syntax of the language ultimately used on the Web. Users can concentrate on the concepts and relationships in the domain and the facts about them that need to be expressed. For example, for development of ontology of wines, food, and appropriate wine–food combinations, authors can focus on Bordeaux and lamb instead of markup tags and correct syntax (Informatics 2010).

Protégé is highly customizable, which makes its adaptation as an editor for a new language faster than creating a new editor from scratch. The following features make this customization possible (Noy et al. 2001):

- An extensible knowledge model. It is possible to redefine declaratively the representational primitives that system uses.
- A customizable output file format. Protégé components that translate from the Protégé internal representation to a text representation in any formal language can be implemented.
- A customizable user interface. Protégé user-interface components for displaying and acquiring data can be replaced with new components that fit the new language better.
- An extensible architecture that enables integration with other applications. Protégé can be connected directly to external semantic modules, such as specific reasoning engines operating on the models in the new language.

Besides the fact that Protégé is a good, intuitive and widely used tool, it also provides an open-source Java API that enables access the ontological model and use of Protégé Forms from a Java environment.

8.4.1.1   Ontology Modelling in Protégé

Concerning ontologies modeling, there are essentially two ways to do it: the Protégé-Frames editor; and the Protégé OWL editor (Burgos 2011). The first one is based on the concept of knowledge representation which involves creating of ontologies by building conceptualization of concepts and organizing them in a hierarchy. This phase also includes creation of a set of relationships between concepts and creating instances from this concepts. Connections between ontology classes are defined that can be discovered by querying the system. In addition, there are some important characteristics of Protégé-Frames that must be highlighted: it includes a plug-in architecture which can contain some elements such as graphics, sounds, different storage formats as well as additional support tools (Burgos 2011).

Modelling of ontologies in Protégé OWL editor is a second method (Burgos 2011). OWL as one of the current standard for Semantic Web presents a mean for creation all ontology elements: classes, their properties and instances. Moreover this editor can support OWL and RDF ontologies as well as the edition and visualization of its classes, properties and rules. Possibility of integration and execution of reasoners are equally important. Reasoners are pieces of software that enable the inference property to spread its power across the ontology.

8.4.2   NeOnToolkit

The NeOn Toolkit is the ontology engineering environment originally developed as part of the NeOn Project (Suarez-Figueroa et al. 2012). It is an open source multi-platform ontology engineering environment, which provides comprehensive support for the ontology engineering life-cycle. The toolkit is based on the Eclipse platform, a leading development environment, and provides an extensive set of plug-ins (currently 45 plug-ins are available) covering a variety of ontology engineering activities, including annotation and documentation, development, human-ontology interaction, knowledge acquisition, management, modularization and customization, neon plugins, old main page, ontology dynamics, ontology evaluation, ontology matching, reasoning, inference and reuse.

8.4.3   TopBraid Composer

TopBraid Composer is an enterprise-class modelling environment for developing Semantic Web ontologies and building semantic applications (Waldman, n.d.). It combines world's leading semantic web modelling capabilities with the most comprehensive data conversion options and a powerful *Integrated Development Environment* (IDE) for building semantic web and Linked Data applications (Alatrish 2012).

Fully compliant with W3C standards, *TopBraid Composer* offers comprehensive support for developing, managing and testing configurations of knowledge models and their instance knowledge bases. TopBraid Composer is the leading industrial-strength RDF editor and OWL ontology editor, as well as the best SPARQL tool on the market.

Composer comes in 3 editions: Free (limited free version), Standard and Maestro.

As part of TopBraid Suite, Composer incorporates a flexible and extensible framework with a published API for developing semantic client/server or browser-based solutions that can integrate disparate applications and data sources.

8.4.4   *Vitro*

First developed for a research and scholarship portal at Cornell University, Vitro is a general-purpose web-based ontology and instance editor with customizable public browsing (Lowe et al. 2011). Vitro is a Java web application that runs in a Tomcat servlet container. Vitro allows users to:

- create or populate ontologies in OWL format,
- edit ontology instances and relationships,
- build a public web site that will display collected data,
- search and brows data.

8.4.5   *OWLGrEd*

The OWLGrEd Ontology Visualizer is an online tool for visualizing OWL ontologies using a compact UML-based notation (Liepins et al. 2014). OWLGrEd is a free UML style graphical editor for OWL ontologies (Barzdinš et al. 2010). It has additional features for graphical ontology exploration and development, including interoperability with Protégé.

8.4.6   *Knoodl*

Knoodl is a web-based, collaborative ontology editor. Each resource in an ontology has its own web-page which is half structured content from the ontology and half unstructured content in the form of wikitext. Content in Knoodl is organized into Communities, which can be created by any user. Communities have a role-based permissions model. Ontologies can be imported and exported as OWL files, with or without the associated wikitext.

All Knoodl functionality is exposed as Java Beans so users can write their own semantic applications by writing Java Server Pages and Java Script. Future releases will support ontology guided search and a rich application development framework which will give users the ability to design custom forms for entering data or queries and custom views for displaying data.

Many more ontology editor exists in the word, but among them Protégé is proved to be the most popular (Alatrish 2012), and therefore was used for development of Protus ontologies described in this monograph.

8.5   Educational Ontologies

Educational ontologies are different forms of ontologies that are used in systems for semantic learning. It is possible to identify several categories of these ontologies (Aroyo and Mizoguchi 2003; Chen 2009; Mizoguchi et al. 2007):

- domain ontology,
- task ontology,
- teaching strategy ontology,
- learner model ontology,
- interface ontology,
- communication ontology,
- educational service ontology.

These categories will be presented in more details in subsequent sections.

8.5.1   Domain Ontology

Systems for semantic learning cannot achieve their goals without domain ontology that describes the basic theoretical concepts and relations in the area that is being taught. Domain ontology is not only important for educational purposes but for all Semantic Web applications. Since the goal of systems for semantic learning is to track progress of learners' mastering specific domain, the course authors should present knowledge with this type of ontologies.

There are two major types of domain knowledge—subject domain and structure, which leads to two types of ontologies (Dicheva et al. 2005):

- Domain ontology. A domain ontology represents the basic concepts of the domain under consideration along with their interrelations and basic properties.
- Structure ontology. A structure ontology defines the logical structure of the content. It is generally subjective and depends greatly on the goals of the ontology application. It typically represents hierarchical and navigational relationships.

While a domain ontology can be used as a mechanism for establishing a shared understanding of a specific domain, a structure ontology enforces a disciplined approach to authoring, which is especially important in collaborative and distributed authoring.

8.5.2 Task Ontology

This ontology complements the domain ontology by representing the semantics of the problem being solved. Concepts and relations that are included in this ontology presents a link to the problem types, structures, areas, activities and steps that learner must follow in the problem solving process. For example, task ontology in educational applications may include concepts such as: problem, scenario, question, answer, guide, hint, exercise, explanation, simulation, etc. Task ontology in semantic learning systems formalize tasks and activities of all relevant participants in the system (learners, lecturers, authors), therefore this ontology could also be called *Instructional design ontology, Training ontology, Authoring task ontology*, etc.

The term *task ontology* can be interpreted in two ways (Ikeda et al. 1998):

- task-subtask decomposition together with task categorization such as diagnosis, scheduling, design, etc.
- an ontology for specifying problem solving processes.

The latter shares the word usage with *domain ontology* which means an ontology of a domain and specifies concepts and relations appearing in a domain of interest.

The advantages of the integration of task ontology into educational systems are as follows (Ikeda et al. 1998):

- Task ontology provides human-friendly primitives in terms of which users can easily describe their own problem solving processes (descriptiveness, readability).
- The system can simulate the problem solving processes at the conceptual level and present users with the execution process in terms of conceptual level primitives (conceptual level operations).
- The system translates problem solving knowledge into symbol level code (symbol level operations).

8.5.3 Teaching Strategy Ontology

This ontology provide instructors and authors with capabilities for modelling learning activities, and define knowledge and principles that underpin the pedagogical actions (Aroyo and Mizoguchi 2003). For example, teaching strategy ontology can define a series of actions that are carried out when a learner makes a mistake, or it can define the behaviour of a system that encourages learners to use alternative solutions.

The goal of teaching strategy ontology is to provide the author with a facility to model the author's teaching experiences (Chen et al. 1998). According to each learner's specific error, the author can represent an appropriate teaching strategy with the use of such an ontology.

8.5.4   Learner Model Ontology

Designers of a semantic learning system use concepts from the learner model ontology to build a learner model. These ontologies and their corresponding learner models are the most important element of adaptive system behaviour. Content of learner model ontology mostly depends on the application in which it is implemented. More specifically, the learner model ontology should collect the objective and subjective data about learner. Data on learners' progress and learning history should be recorded, stored and updated for each individual learner.

Learner model ontology helps the author to represent a suitable learner model mechanism so that the intelligent tutoring system can behave adaptively to the learner's understanding state (Mizoguchi 2004). It facilitates to build learner models in intelligent training systems.

8.5.5   Interface Ontology

The purpose of this ontology is to define adaptive behaviour of the semantic learning system at the user interface level. Therefore, interface ontology provides explicit modelling of adaptation for learners with different characteristics (Devedzic 2006).

The aim of this ontology is to help the author define intelligent tutoring system interface in his own style and to make the user interface adaptive to different learners (Chen et al. 1998).

Interface ontology contains definitions of the user interface segments that will be connected to adaptation rules for displaying/removal of user interface elements, based on the data from the task ontology. In this way it is possible to define parts of the interface that will display the introductory remarks, explanations, examples, navigating options, execution of specific functionalities, etc. Then, for example, in order to customize the user interface for learner with global learning style, it is necessary to display navigational elements for that learner, that is, all elements that belong to the specific class (that represent navigational elements) will be presented.

8.5.6   Communication Ontology

Different semantic learning systems, pedagogical agents, educational servers and educational services communicate with each other using messaging (Bittencourt et al. 2009). Communication ontology defines the semantics of the message content, and creates glossary of terms used in messages.

8.5.7   *Educational Service Ontology*

This ontology provides tools for creating computer-readable descriptions of services, defines consequences of their use and explicitly represents the service logic (Paolucci and Sycara 2003). Educational services have their own characteristics, capabilities, interfaces and results, and they all must be recorded in unambiguous, computer understandable form to allow pedagogical agents to recognize them and automatically execute.

8.6   Adaptation Rules

Although ontologies have a set of basic implicit reasoning mechanisms derived from the description logic which they are typically based on (such as classification, relations, instance checking, etc.), they need rules to make further inferences and to express relations that cannot be represented by ontological reasoning (e.g., in a learning domain it could be necessary to express the fact that a topic A is a prerequisite of topic B to make the right suggestions to the learner). Thus, ontologies require a rule system to derive/use further information that cannot be captured by them, and rule systems require ontologies in order to have a shared definition of the concepts and relations mentioned in the rules. Rules also allow adding expressiveness to the representation formalism, reasoning on the instances, and they can be orthogonal to the description logic on which ontologies are based on (Henze et al. 2004).

There are different types of adaptation rules depending on the knowledge they store:

- **Decision rules**. These rules are used to generate the feedback based on user actions. Decision rules are executed based on defined conditions. When the conditions are fulfilled, the system makes decisions on further action based on corresponding rules. Most often, these actions represent a selection of educational material that will be displayed to learner, selection of recommended pages, choice of navigation, etc.
- **Association rules**. The objective of association rules (Romero et al. 2004) is to look for relationships among attributes types in databases, taking place in the antecedent and consequent of the rules. Association rules are typically used in e-commerce to model the clients' preferences and purchases. These rules have the format: IF "user acquires the product A" THEN "user also acquires the product B" with values of support and confidence (Srikant and Agrawal 1997)

greater than a user-specified minimum threshold. In the more general form of
these rules, the rule antecedent and consequent can present more than one
condition. The confidence of the rule is the percentage of transactions that
contain the consequent among transactions that contain the antecedent. The
support of the rule is the percentage of transactions that contain both antecedent
and consequent among all transactions in the data set (Romero et al. 2004).

- **Classification rules**. The objective of classification rules (Romero et al. 2004) is
  to obtain knowledge in order to create a classification system (similar to a
  classification tree). In the antecedent of the rule there are some requirements (in
  form of conditions) that should match a certain object so that it can be considered to belong to the class that identifies the consequent of the rule. From a
  syntactic point of view, the main difference with association rules is that they
  have a single condition in the consequent which is the class identifier name
  (Tran et al. 2006).
- **Prediction rules**. The objective of prediction rules is to predict an objective
  attribute depending on the values of another group of attributes. Its syntax is
  similar to classification rules that have only one condition in the consequent, but
  now it is similar to any other condition (Romero et al. 2005).
- **Causal rules**. These rules are temporal. Causal relationships do not only indicate that the variables are related (associated) in general, more importantly they
  show how the variations of one variable cause changes of other variable.
  Therefore causality is more useful for prediction and reasoning (Li et al. 2013).
- **Optimization rules**. These rules define the actions that are performed in order to
  optimize functionalities, learning process, navigation, code, etc.

Data processing mechanisms integrated in Semantic Web languages restrict the
use of ontologies. Therefore, various attempts to formalize the logical layer
ontologies are present. Semantic Web Rule Language (Horrocks et al. 2004) is
presented as an important step in this direction, which upgrades the previous work
on a development of RuleML languages. The presence of a standardized language
for defining rules allows simultaneous use of ontologies and adaptation rules in
order to improve Semantic Web applications.

Details about adaptation rules, their types, use and implementation process will
be presented in Sect. 9.3.

8.6.1 Semantic Web Rule Language (SWRL)

SWRL (Sicilia et al. 2011) is a language specifically targeted to introduce inference
rules in knowledge models represented in OWL. Semantic Web Rule Language is
probably the most popular formalism in Web community for expressing knowledge
in the form of rules. Specifically, SWRL is based on a combination of Web
Ontology Language (Bechhofer 2009) and Rule Markup Language (Paschke and

Boley 2010) and has been proposed as a W3C candidate standard for formalizing the expression of rules in Web context.

Rules are represented as an implication between antecedent (body) and consequent (head). The intended meaning can be read as "whenever the conditions specified in the antecedent hold, then the conditions specified in the consequent must also hold".

The main advantage of SWRL is the simplicity it offers, while extending the expressiveness of OWL. Another benefit of SWRL is its compatibility with OWL syntax and semantics, since they are both combined in the same logical language.

It is worthy to mention that most of the existing rule-based applications for the Web have adopted SWRL approach in order to express rules. SWRL is neither a highly expressive language (e.g., no negation is available) nor a decidable one, but it remains simple (Papataxiarhis et al. 2009).

Adaptation rules defined in SWRL are in the form:

$$Antecedent \implies consequent.$$

These rules consist of an antecedent (body) and a consequent (head), each of which consists of a (possibly empty) set of atoms. Informally, meaning of the rule is: if the antecedent holds (is *true*), then the consequent must also hold. An empty antecedent is treated as trivially holding (true), and an empty consequent is treated as trivially not holding (false). Rules with an empty antecedent can thus be used to provide unconditional facts; however such unconditional facts are better stated in OWL itself, i.e., without the use of the rule construct.

A detailed description of adaptation rules used for creation of general tutoring system model will be shown in Sect. 9.3 named *Adaptation rules*.

8.6.2   Jess

Usually, the SWRL rules are translated into existing rule systems (e.g., Jess) that handle the reasoning tasks partially, since they are not aimed to manage knowledge expressed in terms of first-order logic or its subsets (Papataxiarhis et al. 2009).

Jess (*Java Expert System Shell*) is a Java framework for editing and applying rules, since it contains a scripting environment and a rule engine, as well (Friedman-Hill 1997). Recently, the evolution of rule technologies on the Web has led Jess to rebound its practical value in the community of Web developers. Moreover, the fact that Jess is a Java-based system facilitates its integration with a number of Web programming paradigms like Java servlets or applets. Finally, Jess was accessed via *SWRL-Jess Bridge*. Jess is also a rule-based inference engine that can support RDF, OWL and SWRL inference. For this inference it must use SWRLTab, which is one of the OWL Plug-ins for Protégé (Lee et al. 2005).

Defined SWRL rules can be executed using the Jess application programming interface that enables creation of the new instance of *Jess rule engine*. An instance of this class loads all defined rules, check the terms defined in them and starts executing rules if all conditions are met. After execution of the rule, the inferred knowledge can be written back to the ontology repository and update the knowledge base.

8.7   Architecture of Semantic E-Learning Systems

The first necessary precondition for the development of semantic e-learning systems is an adequate development of the Semantic Web. It is necessary that the Semantic Web is sufficiently widespread as the Web itself. Naturally, it is a long and systematic process. Parallel with the development of the Semantic Web, it is necessary that more and more educational content exists in the semantic environment. It involves the development of a large number of ontologies, description of educational content using these ontologies, the development of special education services of the Semantic Web, as well as the development of Semantic Web languages, tools and related technologies.

Figure 8.6 shows the environment required for the storage of educational materials for teaching, testing, evaluation and other educational activities in the Semantic Web (Devedžić 2004; Mustapacsa et al. 2010; Shah 2012). This environment is a generalization of the virtual classrooms architecture. Educational material is available through a variety of educational servers and special Web applications that are responsible for the management, administration and access to material from physical servers. Learners, lecturers and course authors access educational material from the client side. The educational content is an arbitrary material, pedagogically organized and structured in such a way that interested learners can use it to become familiar with the domain knowledge, in order to

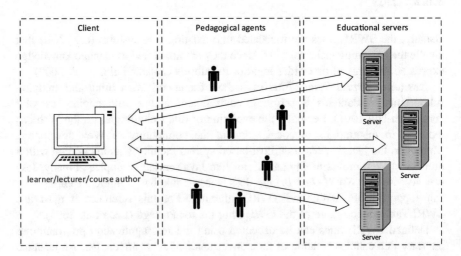

Fig. 8.6 The learning environment in the Semantic Web

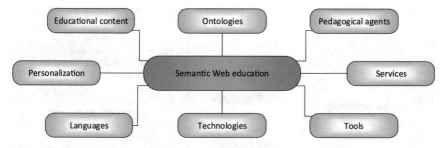

Fig. 8.7 Concepts of semantic web learning systems

understand it and develop the ability to solve appropriate problems. Intelligent pedagogical agents offer the necessary information, knowledge and content flow between clients and servers.

Figure 8.7 shows the basic concepts of machine learning based on the semantic Web. Their brief specifications are (Devedzic 2006):

- **The authors** are preparing **educational content** in the form of multimedia teaching material, such as questionnaires, tests, exercises, simulations, etc. This content is usually structured as a coherent unit of learning, such as lessons, chapters or tests.
- **Ontologies** represent basic knowledge (such as knowledge of the domain or pedagogical knowledge) by defining the terminology, concepts, relationships, hierarchies of concepts and constraints. It allows sharing and reuse of educational content and cooperation between different educational applications.
- **Pedagogical agents** help in locating, searching, selecting, organizing and integrating teaching materials from different educational servers. They are also used to support individual and group learning and to support cognitive processes of learners.
- **Learners** are always interested in **personalized learning**, as all people have their own personal habits, approaches, goals, desires and pace of learning.
- Different **natural, visual, and representation languages** are used to decode and display information contained in the learning material. Also, different formal languages can be used to develop educational content for display of ontologies and educational services. Different languages are used to define the communication between different pedagogical agents.
- Although technology is not the ultimate goal of the e-learning systems, its use is key development factor. **Trends and polarity** of current technological support must not be ignored when e-learning systems are developed.
- **Various tools for learning, teaching and creating courses** often come as powerful integrated Web software applications. There are also many other software tools of different possibilities for the preparation of educational content.

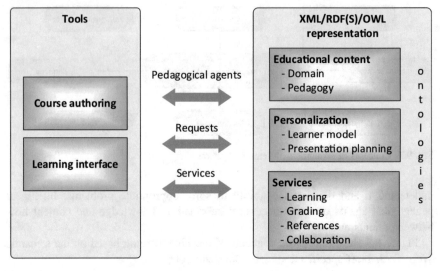

Fig. 8.8 Educational server model (Shah 2012)

- **Semantic Web Services** are used to offer teachers, learners, and course authors access to educational content from different domains of interest. They are usually associated with educational servers and can support a number of different educational activities.

Figure 8.8 shows the general model of educational server for semantic learning systems (Devedžic 2004; Shah 2012). Educational server should use intelligent technologies to personalize learning materials. Server should possess personalization planner that helps intelligent tutor to choose, prepare and adapt material from the domain that will be displayed for learner in accordance to his/her learning manner and style.

E-learning system gradually re-builds the learner model during sessions, in order to keep track of the learner's actions and his/her progress, to detect and correct his/her errors and possibly to redirect the session accordingly. At the end of the session, all data about learner are recorded in learner model. The learner model is then used along to other information and knowledge to initialize the next session for the same learner.

From the teachers' perspective, educational server allows them to access, view and search collections of learning objects and access educational materials. Server also offers the possibility to monitor different classes of learners and provide access to the learner model, both during and after the presentation sessions. Server itself may provide the tools to perform the basic course activities or can use client tools for connection to server and use educational services that server provides.

From the perspective of a course author, educational servers extend authoring tools. With the help of authoring tools, domain authors gain access to the various ontologies stored in the educational server libraries. Authors can define their own ontology and publish them in the library. They also can reuse existing ontologies from the library and perform the necessary changes to them. Authors can create, maintain, update and delete learning objects and courses, change instructional design, approach or update learner model, etc.

Learners must be provided with individual path through educational material and the possibility of cooperation with other learners. Important goals of educational systems are ease of access to the learning material and content filtering for individual learners. Ontologies are used to provide personalization and semantic search and integration of content from different sources.

In order to personalize the learning process and adapt content to each individual learner, e-learning systems must use strategies that will meet the needs of individual learners. Also, these systems must use different technologies for adaptation of user interface and teaching materials based on the needs of learners (Devedzic 2006). The process of personalisation can include adaptation of content, learning process, feedback or navigation. As a result of such adaptation, the same lesson can be presented to different learners in a completely different way.

Consider for example, the case of system for electronic learning of Java programming language, where learners visit series of lessons about Java programming language syntax, and at the same time test their newly acquired knowledge (Vesin et al. 2012). If learner wants to learn the syntax of the loop statements, the system would first display educational materials that explain the syntax rules of these statements, and then present a range of appropriate, illustrative solved examples. If another learner wants to learn the practical application of these commands, the system could test learners' knowledge of the basic concepts of the Java language syntax (for example: basic expressions or declaration of variables), and only after that, system would present details about syntax and use of loop statements. At the end of the process, system could eventually test learners' new gained knowledge from current lesson.

References

Alatrish, E. S. (2012). Comparison of ontology editors. *E-RAF Journal on Computing, 4*, 23–38.

Alsultanny, Y. A. (2006). e-Learning system overview based on semantic web. *The Electronic Journal of E-Learning, 4*(2), 111–118.

Aroyo, L., & Mizoguchi, R. (2003). Authoring Support framework for intelligent educational systems. In *AIED-2003* (pp. 362–364).

Aroyo, L., & Dicheva, D. (2004). The new challenges for e-learning: The Educational Semantic Web. *Educational Technology and Society*.

Babu, P. B., & Krishnamurthy, M. (2013). Library automation to resource discovery: a review of emerging challenges. *The Electronic Library, 31*(4), 433–451. doi:10.1108/EL-11-2011-0159

Barzdinš, J., Barzdinš, G., Čerans, K., Liepinš, R., & Sprogis, A. (2010). OWLGrEd: A UML style graphical editor for OWL. In *CEUR Workshop Proceedings* (Vol. 596, pp. 23–28). doi:10.1007/978-3-642-16101-8_9

Bechhofer, S. (2009). OWL: Web ontology language. In *Encyclopedia of Database Systems* (pp. 2008–2009). Berlin: Springer.

Berners-Lee, T. (2000). Semantic web-xml2000. http://www.w3.org/2000/Talks/1206-xml2k-Tbl/

Berners-Lee, T., Hendler, J., Lassila, O., et al. (2001). The semantic web. *Scientific American, 284* (5), 28–37.

Bittencourt, I. I., Costa, E., Silva, M., & Soares, E. (2009). A computational model for developing semantic web-based educational systems. *Knowledge-Based Systems, 22*(4), 302–315. doi:10. 1016/j.knosys.2009.02.012

Bray, T., Paoli, J., Sperberg-McQueen, C. M., Maler, E., & Yergeau, F. (1998). Extensible markup language (XML). *World Wide Web Consortium Recommendation REC-Xml-19980210*. http:// www.w3.org/TR/1998/REC-Xml-19980210, 16.

Breslin, J. G., Passant, A., & Vrandečić, D. (2011). Social semantic web. In *Handbook of Semantic Web Technologies* (pp. 467–506). Berlin: Springer.

Brickley, D., & Guha, R. V. (2008). RDF Schema 1.1—W3C Recommendation. doi:10.1016/ B978-0-12-373556-0.00006-X.

Burgos, J. L. M. (2011). Semantic web standards. *SNET Computer Engineering*. Retrieved from http://www.pdffiller.com/948565-semantic-web-standards_burgos-Semantic-Web-Standards— SNET-Various-Fillable-Forms-snet-tu-berlin

Cardoso, J. (2007). The semantic web vision: Where are we? *IEEE Intelligent Systems, 22*(5), 84– 88. doi:10.1109/MIS.2007.4338499

Chandrasekaran, B., Josephson, J. R., & Benjamins, V. R. (1999). What are ontologies, and why do we need them? *IEEE Intelligent Systems, 1*, 20–26.

Chen, C.-M. (2009). Ontology-based concept map for planning a personalised learning path. *British Journal of Educational Technology, 40*(6), 1028–1058.

Chen, W., Hayashi, Y., Jin, L., Ikeda, M., & Riichiro Mizoguchi. (1998). An ontology-based intelligent authoring tool. In *Proceedings of the Sixth International Conference on Computers in Education* (pp. 41–49).

d'Aquin, M., Motta, E., Sabou, M., Angeletou, S., Gridinoc, L., Lopez, V., & Guidi, D. (2008). Toward a new generation of semantic web applications. *IEEE Intelligent Systems, 23*(3). doi:10.1109/MIS.2008.54.

Devedzic, V. (2006). *Semantic web and education. Book* (Vol. 11). doi:10.1007/978-0-387-35417-0.

Devedžić, V. (2004). Web intelligence and artificial intelligence in education. *Educational Technology & Society, 7*(4), 29–39.

Devedzic, V., & Harrer, A. (2005). Software patterns in ITS architectures. *International Journal of Artificial Intelligence in Education, 15*(2), 63–94.

Dicheva, D., Sosnovsky, S., Gavrilova, T., & Brusilovsky, P. (2005). Ontological web portal for educational ontologies. In *SW-EL'05: Applications of Semantic Web Technologies for E-Learning* (p. 19).

Dutta, B. (2006). *Semantic web based e-learning*. Bangalore: Documentation Research and Training Centre Indian Statistical Institute.

Fensel, D., & Musen, M. A. (2001). The semantic web: A brain for humankind. *IEEE Intelligent Systems, 16*(2), 24–25.

Friedman-Hill, E. J., & others. (1997). Jess, the java expert system shell. *Distributed Computing Systems, Sandia National Laboratories, USA*.

Gascueña, J. M., Fernandez-Caballero, A., & Gonzalez, P. (2006). Domain ontology for personalized e-learning in educational systems. In *ICALT* (pp. 456–458).

Gilchrist, A. (2000). The well-connected community: Networking to the edge of chaos. *Community Development Journal, 35*(3), 264–275. doi:10.1093/cdj/35.3.264

Gómez-Pérez, A., & Corcho, O. (2002). Ontology languages for the semantic web. *IEEE Intelligent Systems and Their Applications, 17*(1), 54–60. doi:10.1109/5254.988453.

Gruber, T. R. (1995). Toward principles for the design of ontologies used for knowledge sharing? *International Journal of Human-Computer Studies, 43*(5), 907–928.

Gruber, T. (2008). Collective knowledge systems: Where the social web meets the semantic web. *Journal of Web Semantics, 6*(1), 4–13. doi:10.1016/j.websem.2007.11.011

Guarino, N., & Welty, C. (2000). Identity, unity, and individuality: towards a formal toolkit for ontological analysis. In *ECAI 2000* (pp. 219–223).

Henze, N., Dolog, P., & Nejdl, W. (2004). Reasoning and ontologies for personalized e-learning in the semantic web. *Educational Technology & Society, 7*(4), 82–97.

Hodge, G. (2000). *Systems of Knowledge Organization for Digital Libraries: Beyond Traditional Authority Files. Knowledge Organization.*

Horrocks, I. (2008). Ontologies and the semantic web. *Communications of the ACM.* DOI 10. 1145/1409360.1409377.

Horrocks, I., Parsia, B., Patel-Schneider, P., & Hendler, J. (2005). Semantic web architecture: Stack or two towers? In *Principles and practice of semantic web reasoning* (pp. 37–41). Berlin: Springer.

Horrocks, I., Patel-Schneider, P. F., Boley, H., Tabet, S., Grosof, B., Dean, M., & others. (2004). SWRL: A semantic web rule language combining OWL and RuleML. *W3C Member Submission, 21,* 79.

Ikeda, M., Seta, K., Kakusho, O., & Mizoguchi, R. (1998). Task ontology: Ontology for building conceptual problem solving models. In *Proceedings of ECAI98* (pp. 126–133).

Informatics, S. M. (2010). The Protégé ontology editor and knowledge acquisition system welcome to protég. *Knowledge Acquisition,* 2010–2010.

Jovanovic, J., Rao, S., Gasevic, D., Hatala, M., & Devedzic, V. (2007). An Ontological Framework for Educational Feedback.

Klein, M. (2001). XML, RDF, and relatives. *IEEE Intelligent Systems, 2,* 26–28.

Knublauch, H. (2004). Ontology-driven software development in the context of the semantic web: An example scenario with protege/OWL. In *1st International Workshop on the ModelDriven Semantic Web MDSW2004.*

Lee, M.-C., Ye, D. Y., & Wang, T. I. (2005). Java learning object ontology. In *Advanced Learning Technologies, 2005. ICALT 2005. Fifth IEEE International Conference on* (pp. 538–542).

Li, J., Le, T. D., Liu, L., Liu, J., Jin, Z., & Sun, B. (2013). Mining causal association rules. In *2013 IEEE 13th International Conference on Data Mining Workshops (ICDMW)* (pp. 114–123).

Lowe, B., Caruso, B., Cappadona, N., Worthington, M., Mitchell, S., & Corson-Rikert, J. (2011). The vitro integrated ontology editor and semantic web application. In *ICBO.*

Mizoguchi, R. (2004). Tutorial on ontological engineering: Part 02: Ontology development, tools and languages. *New Generation Computing, 22,* 61–96.

Mizoguchi, R., & Bourdeau, J. (2015). Using Ontological Engineering to Overcome AI-ED Problems: Contribution, Impact and Perspectives. *International Journal of Artificial Intelligence in Education,* 1–16.

Mizoguchi, R., Hayashi, Y., & Bourdeau, J. (2007). Inside theory-aware and standards-compliant authoring system. In *SW-EL'07* (18 p.).

Mustapa\csa, O., Karahoca, D., Karahoca, A., Yücel, A., & Uzunboylu, H. (2010). Implementation of semantic web mining on e-learning. *Procedia-Social and Behavioral Sciences, 2*(2), 5820–5823.

Noy, N. F., Sintek, M., Decker, S., Crubézy, M., Fergerson, R. W., & Musen, M. A. (2001). Creating semantic web contents with protégé-2000. *IEEE Intelligent Systems and Their Applications, 16*(2), 60–71. doi:10.1109/5254.920601

Paolucci, M., & Sycara, K. (2003). Autonomous semantic web services. *IEEE Internet Computing, 7*(5), 34–41. doi:10.1109/MIC.2003.1232516

Papataxiarhis, V., Tsetsos, V., Karali, I., Stamatopoulos, P., & Hadjiefthymiades, S. (2009). Developing rule-based applications for the Web: Methodologies and Tools. In *Handbook of research on emerging rule-based languages and technologies: Open solutions and approaches.*

Paschke, A., & Boley, H. (2010). Rule markup languages and semantic web rule languages. *Rule Markup Languages and Semantic Web Rule Languages,* 1–24. doi:10.4018/978-1-60566-402-6.ch001.

Protégé. (2011). *The Protégé ontology editor. Financial executive* (Vol. 19). doi:10.5121/ijait. 2011.1401.

Romero, C., Ventura, S., & De Bra, P. (2004). Knowledge discovery with genetic programming for providing feedback to courseware authors. *User Modelling and User-Adapted Interaction, 14*(5), 425–464. doi:10.1007/s11257-004-7961-2

Romero, C., Ventura, S., Hervas, C., & Gonzalez, P. (2005). Rule discovery in web-based educational systems using grammar-based genetic programming. *Data Mining VI: Data Mining, Text Mining and Their Business Applications*, 205–214. doi:10.2495/DATA050211

Shadbolt, N., Hall, W., & Berners-Lee, T. (2006). The semantic web revisited. *IEEE Intelligent Systems*. http://doi.org/10.1109/MIS.2006.62.

Shah, N. K. (2012). E-learning and semantic web. *International Journal of E-Education, E-Business, E-Management and E-Learning, 2*(2), 113.

Sheth, A., Ramakrishnan, C., & Thomas, C. (2005). Semantics for the semantic web: The implicit, the formal and the powerful. *International Journal on Semantic Web and Information Systems (IJSWIS), 1*(1), 1–18.

Sicilia, M.-Á., Lytras, M. D., Sánchez-Alonso, S., García-Barriocanal, E., & Zapata-Ros, M. (2011). Modeling instructional-design theories with ontologies: Using methods to check, generate and search learning designs. *Computers in Human Behavior, 27*(4), 1389–1398.

Sivashanmugam, K., Sivashanmugam, K., Verma, K., Verma, K., Sheth, A., Sheth, et al. (2003). Adding Semantics to Web Services Standards. In *Proceedings of the International Conference on Web Services* (pp. 395–401).

Srikant, R., & Agrawal, R. (1997). Mining generalized association rules. *Future Generation Computer Systems*. doi:10.1016/S0167-739X(97)00019-8.

Staab, S., Studer, R., Schnurr, H. P., & Sure, Y. (2001). Knowledge processes and ontologies. *IEEE Intelligent Systems and Their Applications, 16*(1), 26–34. doi:10.1109/5254.912382

Suarez-Figueroa, M. C., Gomez-Perez, A., & Fernandez-Lopez, M. (2012). The NeOn methodology for ontology engineering. In *Ontology Engineering in a Networked World* (pp. 9–34). doi:10.1007/978-3-642-24794-1

Swartout, W., & Tate, A. (1999). Guest editors' introduction: Ontologies. *IEEE Intelligent Systems, 1*, 18–19.

Tran, T., Cimiano, P., & Ankolekar, A. (2006). Rules for an ontology-based approach to adaptation. In *Proceedings—SMAP 2006: 1st International Workshop on Semantic Media Adaptation and Personalization* (pp. 49–54). doi:10.1109/SMAP.2006.31.

Vesin, B., Ivanović, M., Klašnja-Milićević, A., & Budimac, Z. (2012). Protus 2.0: Ontology-based semantic recommendation in programming tutoring system. *Expert Systems with Applications, 39*, 12229–12246. doi:10.1016/j.eswa.2012.04.052

Chapter 9
Design and Implementation of General Tutoring System Model

Abstract Regardless of used methodology, central problem in creating web-based educational systems and taking benefits from their wide use is the fact that the current approaches are rather inflexible and inefficient. Design of such systems must be directed to allow reuse or sharing of content, knowledge, and functional components of those systems. According to techniques and methodologies presented in previous chapters, it is possible to develop modern personalized educational system and fully use benefits that Semantic Web technologies offer. In this chapter, general tutoring model is presented that allows building the personalized courses from various domains. This chapter presents architecture of a general tutoring system whose components are modelled and implemented using Semantic Web technologies. Presented tutoring system framework offers options to build, organize and update specific learning resources (educational materials, learner profiles, learning path through materials, and so on.).

Regardless of used methodology, central problem in creating web-based educational systems and taking benefits from their wide use is the fact that the current approaches rather inflexible and inefficient (Aroyo and Dicheva 2004). Design of such systems must be directed to allow reuse or sharing of content, knowledge, and functional components of those systems. According to previous presented techniques and methodologies, it is possible to developed modern personalized educational system and fully use benefits that semantic web technologies offer. In next chapter, general tutoring model will be presented. It allows building personalized courses from various domains.

Four basic components are necessary for building the semantic web framework for tutoring system model (Borland 2007):

- **Machine readable markup for web content**. XML (eXtensible Markup Language) is used for creation of *self-describing* documents. New markup languages will be necessary to define semantics of a web content.
- **Tools that can read and index semantic markup**. The semantic web rely on distributed resources indexed by a centralized knowledge base.

© Springer International Publishing Switzerland 2017
A. Klašnja-Milićević et al., *E-Learning Systems*,
Intelligent Systems Reference Library 112, DOI 10.1007/978-3-319-41163-7_9

- **Development of tools and services that process semantic information**. This component can be viewed as service ready to communicate with intelligent agents, not just humans.
- **Semantic agents** with capabilities to reason and to support decision making.

This chapter presents architecture of a general tutoring system model that can be used to build courses from different domains. All elements of the tutoring system are modelled and implemented using Semantic Web technologies. Presented tutoring system framework offers options to build, organize and update specific learning resources (educational materials, learner profiles, learning path through materials, etc.).

Defined tutoring system model can deploy an unlimited number of personalized courses from different domains and contain formal rules for adapting educational material to each individual learner. The model includes definitions of basic concepts of ontological architecture and specification of the relations between the defined concepts.

Presented architecture extends the usage of Semantic Web concepts, where the representation of each component is made by a specific ontology, allowing a clear separation of the tutoring system components and explicit communication among them. In addition to the ontological system architecture, adaptation rules that allow easy incorporation and modification of personalisation options are defined.

General tutoring system model allows formalization of the educational material and components of an existing tutoring system. Protus 2.1 is a new version of Protus (Vesin et al. 2013). The new version is created by extending the general model proposed in this chapter. Development of teaching materials for Java programming and its integration in the defined tutoring system model will be shown in Chap. 10.

9.1 Architecture of General Tutoring System Model

The main purpose of the general tutoring system model is an adaptation of teaching materials tailored to particular learners based on their knowledge, needs and desires with the help of integrated recommendation systems. Proposed architecture is highly modular and includes five central components: application module, adaptation module, domain module, learner model and session monitor (Fig. 9.1).

The domain module presents storage for all essential learning material, tutorials and tests. It is presented with *the domain ontology* and describes how the information content is structured.

The learner model is a collection of both static and dynamic data about the learner. The system uses that information in order to predict the learner's behaviour, and thereby adapt educational material to his/her individual needs. This model is presented by *Learner model ontology*.

The application module performs the adaptation. To be exact, *the adaptation module* follows the instructional directions specified by *the application module*.

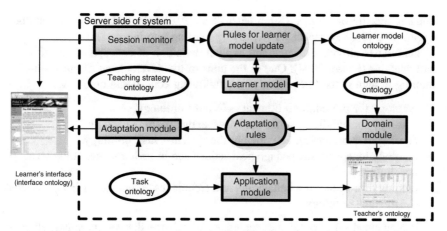

Fig. 9.1 Architecture of general tutoring system model

These two components are separated in order to make adding of a new adaptation functionalities easier.

Within *the session monitor* component, the system gradually re-builds *the learner model* during sessions, in order to keep track of the learner's actions and his/her progress, detect and correct his/her errors and possibly redirect the session accordingly. At the end of every session, *the learner model* is updated and then used along with other information and knowledge to initialize the next session for the same learner. This process is carried out with adaptation rules for learner model update.

The adaptation module contains rules for supporting the adaptive functionality of the system. An example of adaptive functionality is a choice whether certain learner has sufficient knowledge to study a certain lesson/document.

Integration of appropriate educational ontologies and adaptation rules for each component of the system will allow the development of tutoring systems fully supported by the Semantic Web technologies.

Examples of ontologies integrated in general tutoring system model, their use and defined adaptation rules will be presented in Sects. 9.2 and 9.3. General tutoring system model, specifications of tutoring system components, the form of defined ontologies and precise definition of integrated adaptation rules will also be presented.

9.2   System's Ontologies

Over the last decade, special attention has been focused on ontologies and their use in applications in the field of education, knowledge management end data integration. Ontology engineering is a set of actions directed toward the development of domain specific ontologies (Strohmaier et al. 2013). On that basis, the ontology engineering is a key aspect for improving existing tutoring system.

Ontologies provide a formal and explicit definition of concepts, their attributes and relations (Fensel 2004). One of the aims of the development of our Protus 2.1 tutoring system is the integration of the following educational ontologies:

- ontology for presenting a domain—**domain ontology**,
- ontology for building learner model—**learner model ontology**,
- ontology for presenting activities in the system—**task ontology**,
- ontology for specifying pedagogical actions and behaviours—**teaching strategy ontology** and
- ontology for specifying behaviours and techniques at the learner interface level —**interface ontology**.

Implemented set of ontologies facilitates knowledge sharing, its re-use, efficient learner modelling, and scalability of the system. Ontologies are defined using OWL —formal language for creating ontologies (Bechhofer 2009). Ontological architecture of our system is not only used for display of meta-data, but also for creating decisions about the personalization and customization of educational material towards needs of individual learners (Fig. 9.2).

When creating ontologies, it is necessary to use open, standardized languages such as XML, RDF and OWL. They will enable the standardization and formalization of content and enable reuse of system components. Therefore, tutoring system architecture presented in this chapter will enable semantic representation of components and implementation of system's adaptivity (Vesin et al. 2012).

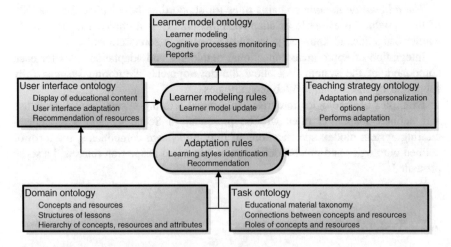

Fig. 9.2 Ontologies and adaptation rules in tutoring system model

9.2.1 Main Components of Ontologies

Ontology is a formal explicit description of concepts in the domain, their attributes and constraints. The basic components of ontologies are (Wand et al. 1999):

- **Classes** (*types, sorts, categories, kinds*). Classes are collections of objects, i.e. concepts.
- **Individuals** (*Instances*). Individuals are concrete objects created from classes.
- **Attributes** (*properties, slots*). Attribute is a directed binary relation that specifies class characteristics. They represent attributes of instances and sometimes act as data values (*Datatype property*) or link to other instances (*Object property*). OWL also has a third type of property—*Annotation properties*. Annotation properties can be used to add information (metadata) to classes, individuals and object/datatype properties. OWL allows classes, properties, individuals and ontology itself to be annotated with various pieces of information/meta-data. These pieces of information may take the form of auditing or editorial information. For example, it could represent details about creation date, author, comments or references to resources such as Web pages etc.
- **Relationships**. Relations represent a way in which certain objects are in relation to other objects.

Classes and a set of individual instances of the classes make ontology knowledge base.

9.2.2 Domain Ontology

One of the main goals of the learning process is to understand and to acquire a body of knowledge for a given domain (Brusilovsky 2004). Domain module presents storage for all essential learning material, tutorials and tests. It describes how the content intended for learning has to be structured. The domain module is structured as a taxonomy of concepts, with attributes and relations connecting them with other concepts, which naturally leads to the idea of using ontologies to represent this knowledge. Ontology for a particular domain of knowledge usually contains (Vesin et al. 2012):

- **Course taxonomy**—contains definitions of types and use of educational materials. Therefore, general tutoring system model offers the possibility for creating educational material for a presentation of theoretical explanations, additional materials, tests, etc.
- **Domain knowledge**—consists of elements that represent concrete concepts (*lessons*). These concepts can be divided into categories, and are specific for each course that was developed with defined general tutoring system model. For

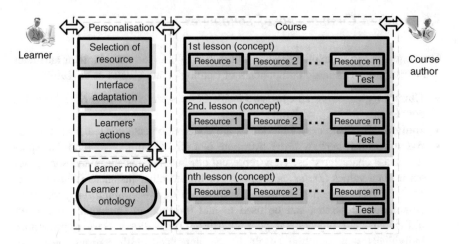

Fig. 9.3 Educational material in general tutoring system model

example, following concepts should be formed in order to implement online programming course: syntax rules, loop statements, branching statements, methods, etc.

Proposed general tutoring system model allows development of courses from different domains. Each course consists of a series of lessons (concepts) showing the individual segments of the domain being taught (Fig. 9.3). Each lesson consists of a series of resources that represent individual parts of the lesson (*introduction*, *explanations*, *examples*, *practice assignments*, etc.). Each lesson is linked to one or more suitable tests to check learners' knowledge. Based on the test results, system determines learners' progress, update the learner model and generate personalization options for learners.

9.2.2.1 Concepts

Ontology class *Concept* is used for storing data about a single lesson. Each concept is presented by an arbitrary number of resources (class *Resources*) in the form of text files, images, illustrations, charts, etc. Each resource is defined by its resource type (class *Resource type*). Therefore, resources can be used to present: theory (class *Theory*), examples (class *Example*), tasks (class *Assignment*), exercise (class *Exercise*), explanations (class *Explanation*), etc. Resources defined in the general tutoring system model are stored in the form of *html* documents.

Datatype properties of classes *Concept* and *Resource* are shown in Table 9.1. Other information in the ontology is stored in the form of relations with the corresponding classes (*Object properties*), which will be shown in Sect. 9.2.3 named *Task ontology*.

Table 9.1 Learner model attributes in classes *Concept* and *Resource*

Concept	
hasId[type:int]	Identification number of concept
hasName[type:string]	Concepts name
Resource	
hasId[type:int]	Identification number of resource
hasName[type:string]	Resource's name
supports[type: string]	Resource's role
isVisited[type: string]	Is resource visited?
isRecommended[type: string]	Is resource recommended?
hasFileType[type: string]	File type
hasRole[type: string]	Resource's role

9.2.2.2 Resources

All concepts must be supported with various types of resources. Types of resources are defined in the resource ontology that is part of the domain ontology. During learning sessions, system selects which of the available resources will be presented depending on generated recommendations for particular learner.

Details about resources are kept in *Resource* class instances. Each instance of the *Resource* class contains basic information on individual resources, which will later be used for the subsequent selection of appropriate resources in the process of personalization. Specific type and role are determined for every resource (Table 9.1).

All resources are grouped according to their type, role and assigned concept and this categorization is the basis of successful recommendations of resources during the personalization process.

Each resource is assigned by the appropriate resource type (class *ResourceType*). Topology of *Resource type* ontology is shown in Fig. 9.4. This ontology contains the types of resources that are required for a specific course. It is possible to define specific resource types depending on structure and particularities of the course being taught. The most general resource type is *DomainResource* (Vesin et al. 2012). *DomainResource* has three subtypes: *CourseMaterial*, *AdditionalMaterial* and *ExaminationMaterial*. Classes *CourseMaterial* and *AdditionalMaterial* represent the theoretical and practical explanations, respectively, that are presented to the learners.

ExaminationMaterial can be further specialized to *Task* and *Exam*. The *Exam* is consisted of various *Tasks*. Since the first developed course material has been for the programming course, most types of tasks are programming oriented. Tasks could include code completion, code correction, listing errors, etc. During the development of courses from other domains (not programming) it is possible to define general type tasks (class type *General*).

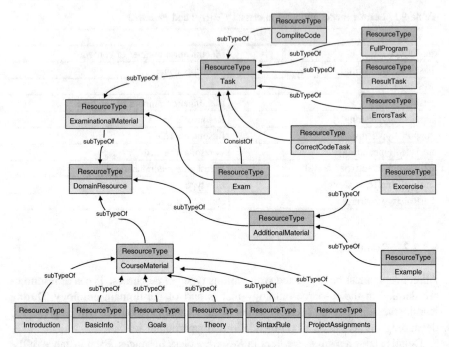

Fig. 9.4 An excerpt of resource ontology of general tutoring system model

The entire educational material of a course (class *CourseMaterial*) can be further subdivided into smaller instructional units. *CourseMaterial* can contain: introduction (presented by *Introduction* class), basic information (class *BasicInfo*), goals (class *Goals*), theory (class *Theory*), various explanations (class *Explanation*) and definitions of project assignments (class *ProjectAssignements*). All these types of materials correspond to the basic types of educational materials of the programming course but can also be used for the more general courses. It is also possible to define new types of materials.

Ontology presented in Fig. 9.4 provides further information for the *Task ontology* and the *Teaching strategy ontology* which will be explained in more details in Sects. 9.2.3 and 9.2.4.

9.2.3   Task Ontology

Task ontology is a vocabulary for domain independently description of a problem solving structure and educational tasks (Vesin et al. 2013). It complements the domain ontology by representing semantic features of the problem-solving (Devedzic 2006). Task ontology specifies domain knowledge by giving roles to each object and presents relations between them. This ontology does not describe

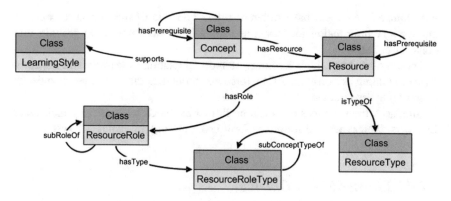

Fig. 9.5 Task ontology in general tutoring system model

the content taught by the learning material. Instead, each class of the ontology stands for a particular instructional role for a learning concept.

Task ontology shows the role of specific resource from domain otology. For example, if resource has *fact* or *definition* role it is used to increase basic knowledge and if its role is *example*, than it is used to increase learner's practical skills.

An excerpt of task ontology of resources in General tutoring system model is depicted in Fig. 9.5. The ontology represents learning material grouped by the resources. The class *Concept* is used to annotate a unit of knowledge which is represented by some *Resource*.

Details about resources are kept in *Resource* class instances. Each instance of *Resource* class contains basic information on individual resources, which are used for the subsequent selection of appropriate resources in the process of personalization. Specific type and role is determined for every resource.

Each concept is presented to a learner using number of resources. In this way, concepts in the general tutoring system model contain: introduction, theory, explanations, examples, tasks, etc. Each of these elements is shown using adequate resources, tor example, HTML or text files, images in jpg format, etc.

Concepts and resources can be arranged by the *hasPrerequisite* property. The *hasPrerequisite* property is proposed for navigational purposes. It allows pointing out concepts/resources that must be mastered before starting to study a certain concept/resource, and the concepts/resource for which it is a prerequisite. Concept/resource will not be covered unless that the prerequisite condition is satisfied. There can be different sequence of concepts/resources that depends on navigational sequence determined for particular learner. SWRL rules update ontologies and implement navigation through the educational materials.

Resources play certain roles in particular resource fragments. For example, some resources represent the crucial information, while the others just represent a mean to provide additional information or a comparison.

In the proposed ontology, we represent these facts by instances of *ResourceRole* class and its two properties: *hasRole* and *supports*. For example, resources like

BasicInfo and *Example* have different roles. The role of the first is to represent introductory information for lesson and the role of the former is to provide additional information.

On the other hand, both resources support adaptation to learner with *Reflective* style of learning (Vesin et al. 2011). Resource properties can be further extended by assigning a *ResourceType*.

Similarly, the resources roles can be further extended by specifying their types. Concepts, their types and resources form task ontology of Protus 2.1 system.

9.2.4 Learner Model Ontology

Building the learner model and tracking related cognitive processes are important aspects in providing personalization. The learner model serves for a representation of information about an individual learner that is essential for an adaptive system to provide the adaptation effect. The system uses that information from learner model in order to predict the learner's behaviour, and thereby adapt to his/her individual needs (Vesin et al. 2012). Data from learner model in Protus 2.1 is classified along three layers as presented in Fig. 9.6.

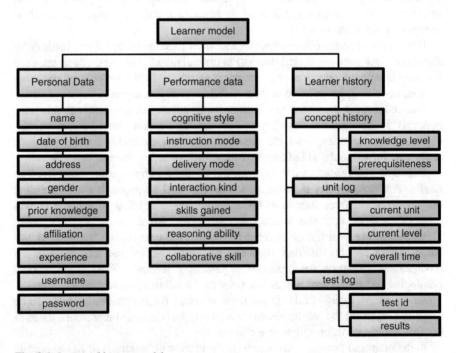

Fig. 9.6 Layers of learner model

- **Objective information**, which includes data supplied directly by the learner like: personal data, previous knowledge, preferences, etc. The learner edits this data during his/her registration on the system.
- **Learner's performance**, which includes data about level of knowledge of the subject domain, his/her misconceptions, progress and the overall performance for particular learner.
- **Learning history**, which includes information about lessons and tests learner has already studied, his/her interaction with system, the assessments (s)he underwent, etc.

The learner model stores personal preferences and information about the learner's mastery of domain concepts (Ullrich 2004). The information is regularly updated according to the learner's interactions with the content and is used by the *Teaching strategy ontology* to draw conclusions and decisions.

Learner model ontology collects information about a learner and his/her actions: learner's username, access time and numerical results of interaction (visited resources, test results, etc.). This ontology is automatically updated whenever a request or action reaches the Web server (for example, visited lesson, resource, calculated test grade, etc.).

The ontology illustrated in Fig. 9.7 offers the opportunity to map all information about the learner, starting from confidential data (like password) to the knowledge evolution history.

Tutoring system model defines two categories of users: teachers (class *Teacher*) and learners (class *Learner*). *Teacher* and *Learner* classes are subclasses of class *User*.

The class *Learner* is built from three components: *Performance*, *PersonalInfo*, and *LearningStyle*. These three classes are related to association through *hasPerformance*, *hasInfo*, and *hasLearningStyle* properties (Fig. 9.7).

Classes *Learner, Performance, PersonalInfo* i *LearningStyle* represent *Learner model ontology*. Each class contains a number of attributes (*Datatype Properties*)

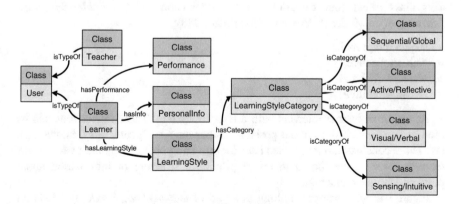

Fig. 9.7 Learner model ontology in general tutoring system model

Table 9.2 Class attributes of learner model ontology

Learner	
id [type:int]	Learner's identification number
PersonalInfo	
learner[type:int]	Learner's identification number
name[type:string]	Learner's name
lastname[type:string]	Learner's last name
gender[type:string]	Learner's gender
address[type:string]	Learner's address
birthDate[type:dateTimeStamp]	Learner's date birth
birthPlace[type:string]	Learner's birth place
previousKnowledge[type:string]	Previous knowledge of the learner
affiliation[type:string]	Learner's affiliation
Performance	
learner[type:int]	Learner's identification number
course[type:int]	Active course's code
percentage[type:double]	Percentage of course
avgGrade[type:double]	Average grade
lesson[type:int]	Last lesson
curLesson[type:int]	Current lesson
resurs[type:int]	Current resource
LearningStyle	
learner[type:int]	Learner's identification number

that are used to store data about each individual learner. Attributes of the classes *Learner*, *Performance*, *PersonalInfo* and *LearningStyle* are shown in Table 9.2. Other information are stored in the ontology using relationships with appropriate classes (*Object properties*), which will be shown later in this chapter.

Class *LearningStyle* represents the preferred learning style for particular learner. This class offers four categories to the dimensions of the *Felder-Silverman Learning Style Model* (Felder and Silverman 1988):

- sequential/global,
- active/reflective,
- visual/verbal and
- sensing/intuitive.

At run time, learner interacts with a tutoring system. These interactions can be used to draw conclusions about possible his/her interests, goals, tasks, knowledge, etc. These conclusions can be used later for providing personalization. *Ontology for learner observations* should therefore provide a structure of information about possible learner interaction.

Figure 9.8 depicts such ontology as a part of *Learner model ontology*. Learner performance is maintained according to a class *Interaction*. *Interaction* is based on

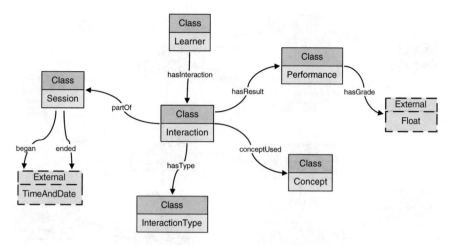

Fig. 9.8 Ontology for learner observation and modeling

actions taken by specific learner, during specific *Session*. *Interaction* implies a *Concept* learned from the experience, which is represented by *conceptUsed* property. *Interaction* has a certain value for *Performance*, which is in this context defined as a floating point number and restricted to the interval from 1 to 5. This ontology is responsible for updating the *Learner model ontology*.

9.2.5 Teaching Strategy Ontology

Authoring of adaptation and personalization is actually authoring of learner models and applying different adaptation strategies and techniques to ensure efficient tailoring of the learning content to the individual learners and their learning style (Aroyo and Mizoguchi 2003; Vesin et al. 2012).

Figure 9.9 shows how the adaptation is carried out by the *Teaching strategy ontology*. The decisions are drawn on the basis of the information contained in the *Condition* class (that is generated by the information about learning style and performance of the learner) as well as teaching goals. Class *AdaptationType* contains information about type of adaptation and are composed of data coming from several other components such as *Learner model ontology*, *Task ontology* and *Domain ontology*.

Personalization presents the choice of the most appropriate learning pattern or resource that will be recommended to the learner. This action depends on many conditions but it implies only one decision. The decision determines what concept and resource the system is going to present to the learner.

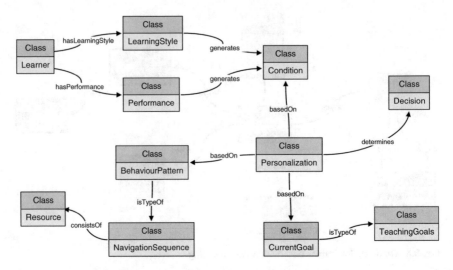

Fig. 9.9 Teaching Strategy ontology of general tutoring system model

9.2.6   Interface Ontology

Interface ontology is result of the final stage of communication among the different components of the architecture. Figure 9.10 shows the steps performed within the *Interface ontology*. System reads a decision from the *Teaching strategy ontology*, and based on that *Decision* it creates *Navigation sequence* of resources recommended for particular learner and generates an interface view to the learner (Vesin et al. 2012).

For example, if *Decision* includes recommending certain resource for learner than presentation will contain *Resource* of specific *Resource* type. On the other hand, if *Decision* contains data about recommended navigational pattern, then Protus 2.1 adds recommended *Resource* to current *Navigational sequence*.

Interface ontology can be used to specify the content of the pages or to standardize content and query vocabulary. This process is implemented with adaptation rules and will be described in Sect. 9.3.

Fig. 9.10 Interface ontology of general tutoring system model

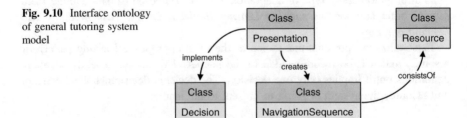

9.3 Adaptation Rules

SWRL rules are one of the most popular forms of knowledge representation, due to its simplicity, comprehensibility and expressive power (Romero et al. 2006). There are different types of adaptation rules depending on the knowledge they store. They are referred to as (as described in Sect. 8.6):

- decision rules,
- association rules,
- classification rules,
- prediction rules,
- causal rules,
- optimization rules, etc.

Rules used in general tutoring system model can be categorized as (Vesin et al. 2012):

- **Learner modelling rules** that add knowledge about a learner. They are necessary for the learner style identification based on observed learning preferences.
- **Adaptation rules** that define the strategies of adaptation, taking into account domain features, system adaptation goals, user features, context and used presentation methods. They are necessary for content adaptation based on identified learning style and/or learning preferences for every particular learner (Popescu et al. 2007).

Adaptation rules for reaching the adaptation goal can be defined taking into account the knowledge domain, learners' current knowledge, his/her preferences and learning style. Also, the definition of adaptation rules requires considering the set of available adaptation methods and techniques (such as hiding text/links, link annotations, presentation methods, altering navigation sequences, etc.) (Carmagnola et al. 2005).

The advantage of presenting teaching strategies in the form of SWRL rules is that these strategies can be explicitly presented, viewed, edited and re-used in other systems.

9.3.1 Syntax of Adaptation Rules

The proposed rules consist of an *antecedent* (body) and a *consequent* (head), each of which consists of a (possibly empty) set of atoms (Horrocks et al. 2004).

Informally, meaning of the rule is: if the antecedent holds (is "true"), then the consequent must also hold. A typical SWRL rule is of the following form:

$$a_1 \wedge a_2 \wedge \ldots \wedge a_n \rightarrow b_1 \wedge b_2 \wedge \ldots \wedge b_m$$

where a_i and b_i are OWL atoms of the following forms:

- Concepts, e.g., C(x), where C is an OWL description, in general, and x is either a variable, an OWL individual (facts about class membership, property values of individuals or facts about individual identity) or a data value.
- Object properties, e.g., P(x,y), where P is an OWL property and x, y are either variables, individuals or data values.
- Datatype properties, e.g., P(x,y), where P is an OWL property, x is variable or individual, while y is a data value.
- B(x1, x2, ..., xn), where B is a built-in relation and x1, x2, ..., xn are either variables, individuals or data values.
- *SameAs(x,y)* or *differentFrom(x,y)* where x, y are either variables, individuals or data values.

An empty antecedent is treated as trivially holding (*true*), and an empty consequent is treated as trivially not holding (*false*). Rules with an empty antecedent can thus be used to provide unconditional facts; however such unconditional facts are better stated in OWL itself, i.e., without the use of the rule construct.

Non-empty antecedents and consequents hold if all of their constituent atoms hold, i.e., they are treated as conjunctions of their atoms.

While the abstract SWRL syntax is consistent with the OWL specification, and is useful for defining XML and RDF serializations, it is rather verbose and not particularly easy to read (Horrocks et al. 2004). In the rest of the section we will, therefore, often use a relatively informal human readable form similar to that used in many published papers.

In this informal syntax, a rule has the form:

$$antecedent \Rightarrow consequent$$

where both antecedent and consequent are conjunctions of atoms written as $a_1 \wedge \ldots \wedge a_n$.

Variables are indicated using the standard convention of prefixing them with a question mark (e.g., ?x). Using this syntax, a rule asserting that the composition of *parent* and *brother* properties implies the *uncle* property would be written:

$$parent(?x, ?y) \wedge brother(?y, ?z) \Rightarrow uncle(?x, ?z)$$

This code defines the rule: if *x*, *y* and *z* are instances of the class *person*, and if relations are defined such that the person *y* is parent of a person *x* and person *y* and *z* are brothers, then automatically person *z* is uncle of persons *x*, i.e. relations *uncle* is defined over these two instances of the class *person*.

In this syntax, built-in relations that are functions can be written in appropriate notation for functions, i.e.:

$$?x = op : numeric-\text{add}(3, ?z)$$

can be written instead of:

$$op : numeric-\text{add}(?x, 3, ?z)$$

During learning sessions, the most important system's task is execution of rules and the subsequent adaptation of the learner interface. Each rule is evaluated and, if all conditions hold, the body (action) of the rule is executed.

Defined rules of presented tutoring system model have general form and could be used in other systems, for development of courses from other domains. System administrators using *Protégé* system can change, delete or add new SWRL rule adaptations which allow adding new personalisation options. *Protégé* offers *SWRLTab* editor for defining SWRL adaptation rules.

The remainder of this section presents the approach of defining rules for the adaptation and personalisation in a general tutoring system model. Typical rules for adaptation of teaching materials to the learner's needs are presented. Adaptation based on the learning style identification by Felder-Silverman model will be presented first, followed by rules for generating recommendations and update of learner model ontology.

9.3.2   Learning Styles Identification

There are over seventy identifiable approaches to investigate and/or describe learning style preferences as presented in the Sect. 3.1. We decided to use in our system one such data collection instrument, called *Index of Learning Styles* (ILS) (Soloman and Felder 2005). The ILS is a 44 question, freely available, multiple-choice learning styles instrument, which assesses variations in individual learning style preferences across four dimensions or domains, including two categories of learners in each dimension:

- **Information Processing domain**: *Active* and *Reflective* learners,
- **Information Perception domain**: *Sensing* and *Intuitive* learners,
- **Information Reception domain**: *Visual* and *Verbal* learners,
- **Information Understanding domain**: *Sequential* and *Global* learners.

In our system, before initial session, and after learning style has been determined by the ILS (questionnaire at the beginning of the course), current learning style category of the particular learner must be written in *Learner model ontology* (Vesin et al. 2012).

For example, if system determines that learner belongs to *Active* category within *Information Processing* domain, Learner model ontology should be updated with that fact with the appropriate rule:

`Learner(?x) ∧ hasLearningStyle(?x,?y) ∧ hasCategory(?y,?z) ∧` `isCategoryOf(?z,active) → hasLearningStyle(?x,active)`	AR1

The meaning of the rule AR1 is: if the *active* learning style is determined for the current learner *x*, then learner model ontology should be updated and active learning style should be assigned to that learner (*active* property). Analog appropriate rules support entering other categories within different learning styles domains.

At the beginning of every session, system requests information about the status of the course from the *Learner model ontology* for the particular learner (Fig. 9.7). This data includes information about the current lesson and the learning style category of learner within each of the four domains of the ILS. Request for appropriate resources which will be presented to the learner, based on this data, is sent to the *Application module*. Further, all activities of learners are monitored, as well as all requests (s)he send to the system.

Personalization actions and presentation of lessons are determined by ILS and learners performance. If conditions (that personalization is based on) are generated by the learning style of the learner (Fig. 9.8) then rule system starts personalization based on appropriate condition. For learner with active learning style, generated condition is *act*, for learner with reflective learning style generated condition is *ref*, etc.

Examples of rules that implement those actions are:

`Learner(?x) ∧ hasLearningStyle(?x,active) ∧ Generates` `(active, ?z) ∧ Condition (?z) → Condition(act)`	AR2
`Learner(?x) ∧ hasLearningStyle(?x, reflective) ∧ Generates` `(reflective, ?z) ∧ Condition (?z) → Condition(ref)`	AR3

The meaning of the rules AR2 and AR3 are: if the currently active learner *x* has a particular learning style (active/reflective) and this style generates a certain condition *z* then an instance of the appropriate class *Condition* should be initialized (*active* style generates a condition *act*, reflective style generates condition *ref*).

Generated conditions are used for determining further actions which will be described in detail later in this section.

After conditions are determined, system makes appropriate decision for presentation based on adaptation type (Vesin et al. 2012).

There are three adaptation types, so far implemented in General tutoring system model: *styleMatch* (matching appropriate learning styles), *adaptInterface* (displaying/hiding interface elements) and *navigation* (altering navigation through course).

Several rules for making decisions about further personalization are:

Personalisation(?p) ∧ basedOn(?p,?c) ∧ Condition(act) ∧ CurrentGoal(?g) isTypeOf(?g, stylematch) → determines(?p, act100)	AR4
Personalisation(?p) ∧ basedOn(?p,?c) ∧ Condition(ref) ∧ CurrentGoal(?g) isTypeOf(?g, stylematch) → determines(?p, ref100)	AR5
Personalisation(?p) ∧ basedOn(?p,?c) ∧ Condition(act) ∧ CurrentGoal(?g) isTypeOf(?g, adaptinterface) → determines(?p, act101)	AR6
Personalisation(?p) ∧ basedOn(?p,?c) ∧ Condition(ref) ∧ CurrentGoal(?g) isTypeOf(?g, adaptinterface) → determines(?p, ref101)	AR7
Personalisation(?p) ∧ basedOn(?p,?c) ∧ Condition(act) ∧ CurrentGoal(?g) isTypeOf(?g, navigation) → determines(?p, act102)	AR8
Personalisation(?p) ∧ basedOn(?p,?c) ∧ Condition(ref) ∧ CurrentGoal(?g) isTypeOf(?g, navigation) → determines(?p, ref102)	AR9

System makes a decision based on the current learning style of the learner and current adaptation type. Therefore, active instances of *Decision* class are determined. *Decision* named *act100* is activated when learner's current learning style is active and *styleMatch* adaptation type needs to be implemented. *Decision* named *ref101* is activated when learner's current learning style is *Reflective* and *adaptInterface* adaptation type needs to be implemented, etc.

Further personalization depends of *Decision* made by previous rules.

9.3.2.1   Adaptation Rules in Information Processing Domain: Active and Reflective Learners

Within *Information Processing* domain it could be distinguished example-oriented learners, named *Reflectors*, and activity-oriented learners, named *Activists* (Kolb 1984).

Active learners tend to retain and understand information best by doing something active with it—discussing or applying it or explaining it to others.

Reflectors are people who tend to collect and analyse data before taking an action. They may be more interested in reviewing other learners' and professional opinions rather than doing real activities.

In General tutoring system model, a learner with the active learning style is shown an activity first, then a theory, explanation and example at the end (Klašnja-Milićević et al. 2011).

For example, in case of a specific learner style, the recommended action would be to present the learner first with the preferred media type and then with the alternative representation types. Several rules that implement presentation of sequences of resources are:

Resource(?x) ∧ isTypeOf(?x, excercise) ∧ Resource(?y) ∧ isTypeOf(?y,example) ∧ Decision(act100) → hasPrerequisite (?y,?x)	AR10
Resource(?x) ∧ isTypeOf(?x, theory) ∧ Resource(?y) ∧ isTypeOf (?y,example) ∧ Decision(act100) → hasPrerequisite(?x,?y)	AR11

Purposes of the previous rules are to define prerequisites among resources. The meaning of the rules AR10 is: if adaptation under the symbol ACT100 is chosen for active learner then resources that present exercises have priority in respect to resources that display examples. Therefore, the rule AR10 defines that exercise is prerequisite for presenting an *example* and meaning of the rule AR11 is to assign higher priority for presenting the theory over examples.

For the learner with the reflective style this order is different—(s)he is shown an *example* first, then an *explanation* and *theory*, and finally (s)he is asked to perform an *activity*.

Example of rules that adapt the order of resources based on reflective learning style are:

Resource(?x) ∧ isTypeOf(?x, excercise) ∧ Resource(?y) ∧ isTypeOf(?y, example) ∧ Decision(ref100) → hasPrerequisite (?x,?y)	AR12
Resource(?x) ∧ isTypeOf(?x, theory) ∧ Resource(?y) ∧ isTypeOf (?y, example) ∧ Decision(ref100) → hasPrerequisite(?y,?x)	AR13

Meanings of these rules (AR12 and AR13) are analogous to previously mentioned (AR10 and AR11). From the learners' point of view, results of the firing this rules are displaying resources in adequate order. Examples of interface for learners with active and reflective learning style are presented in Figs. 9.11 and 9.12, respectively.

Other form of personalization for *Information processing* domain is link annotation. Every resource is given a certain role. For example, if resource has *fact* or *definition* role it is used to increase basic knowledge (preference of the reflective learners).

On the other hand, if role of a resource is to present examples then this resource is used to increase the practical knowledge of learners (for the activists). If certain role is predefined for learners with active learning style, than that resource is recommended to learner. The corresponding rule that implements this activity is:

Learner(?x) ∧ hasLearningStyle(?x,active) ∧ Resource(?r) ∧ ResourceRole(?c) ∧ supports(?r,?c) ∧ supports (?r, active) → isRecommended(?r, true)	AR14

Fig. 9.11 User interface for activists

Fig. 9.12 User interface for reflectors

Fig. 9.13 Recommendation of *Communication* option

The meaning of the previous rule is: if current learner has active learning style, and if resource *r* has a defined role to display teaching materials for *active* learners, then this resource for that learner should be recommended (property *isRecommended* receives *true* value). *isRecommended* is a data valued property atom that consists of an OWL data property (*recommended*).

With rule AR14, recommendation status of that resource is set to true, therefore changes in user interface will be made.

In this case, system seeks resources that are predefined for learners with active learning style. For example, a learner with the active learning style can participate in activities such as quiz, chatting, and discussion options. Therefore, system annotate appropriate link that provides communication options (Fig. 9.13). Similar rules are used for adaptation to other learning styles, too.

9.3.2.2 Adaptation Rules in Information Perception Domain: Sensing and Intuitive Learners

Within *Information Perception* domain sensing learners, named *Sensors*, tend to be patient with details and good at memorizing facts and doing hands-on (laboratory) work. On the other hand *intuitive* learners, named *Intuitors* may be better at grasping new concepts and are often more comfortable than sensing learners with abstractions and mathematical formulations.

Sensors often like solving problems by well-established methods and dislike complications and surprises. On the other hand, *Intuitors* like innovation and dislike repetition. *Sensors* tend to be more practical and careful than Intuitors. *Intuitors* tend to work faster and to be more innovative than Sensors.

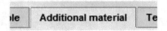

Fig. 9.14 Recommendation of *Additional material* option

Fig. 9.15 Recommendation of *Syntax rules* option

For example, it is assumed that sensing learners will be interested in additional materials, therefore they may click the button for additional material on the interface (Fig. 9.14.) (Klašnja-Milićević et al. 2011). Rules that implement these actions are:

Learner(?x) ∧ hasLearningStyle(?x,sensing) ∧ Resource(?r) ∧ isTypeOf(?r, excercise) ∧ ConceptRole(?c) ∧ supports(?r,?c) ∧ supports (?c, sensing) → isRecommended(?r, true)	AR15
Learner(?x) ∧ hasLearningStyle(?x,sensing) ∧ Resource(?r) ∧ isTypeOf(?r, example) ∧ ConceptRole(?c) ∧ supports(?r,?c) ∧ supports (?c, sensing) → isRecommended(?r, true)	AR16

These adaptation rules set value, of recommendation attribute of specific instances of resources class, to true. In the above case, rules AR15 and AR16 are used to recommend resources to *Sensor* learner. If *Recommended* attribute for resource is set to *true*, it gives an information to *Teaching strategy ontology* that particular resource is appropriate to learner and it could be presented to him/her.

Intuitors are provided with abstract material, formulas and concepts. Adequate explanations are presented in a form of block diagrams or exact syntax rules. Example of rule that implements those actions is:

Learner(?x) ∧ hasLearningStyle(?x,intuitive) ∧ Resource(?r) ∧ isTypeOf(?r, explanation) ∧ ConceptRole(?c) ∧ supports(? r,?c) ∧ supports (?c, intuitive) → isRecommended(?r, true)	AR17

Rule AR17 is used to recommend syntax rule resource to *Intuitive* learner that results in adding appropriate tab in tabbed pane (Fig. 9.15).

9.3.2.3   Adaptation Rules in Information Reception Domain: Visual and Verbal Learners

Within *Information reception* domain, there are two categories of learners: *Visual* and *Verbal*. *Visual* learners remember best what they see—pictures, diagrams, flow charts, time lines, and demonstrations (Klašnja-Milićević et al. 2011).

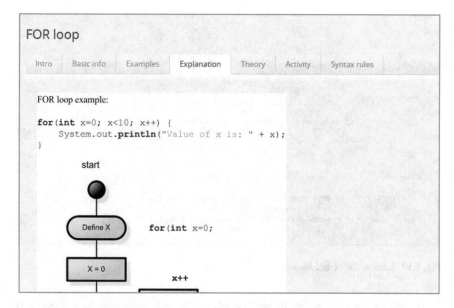

Fig. 9.16 Lesson for visual learners

Verbal learners get more out of words—written and spoken explanations. System recommends appropriate resources with the following rules:

Learner(?x) ∧ hasLearningStyle(?x,visual) ∧ Resource(?r) ∧ ConceptRole(?c) ∧ supports(?r,?c) ∧ supports (?c, visual) → isRecommended(?r, true)	AR18
Learner(?x) ∧ hasLearningStyle(?x,verbal) ∧ Resource(?r) ∧ ConceptRole(?c) ∧ supports(?r,?c) ∧ supports (?c, verbal) → isRecommended(?r, true)	AR19

The rule AR18 is used for recommending resources that contain pictures and diagrams to *Visual* learner (Fig. 9.16) while former rule recommends resource with written explanation to *Verbal* learner (Fig. 9.17). The examples of *verbal* and *visual* presentation of a learning material are given for the Java programming course that is implemented in Protus 2.1. Details of this course will be presented in Chap. 10.

9.3.2.4 Adaptation rules in Information Understanding domain: Sequential and Global learners

Within *Information Understanding* domain, *Sequential* learners tend to follow logical stepwise paths in finding solutions.

On the other hand *Global* learners may be able to solve complex problems quickly or put things together in novel ways once they have grasped the big picture, but they may have difficulty explaining how they did it.

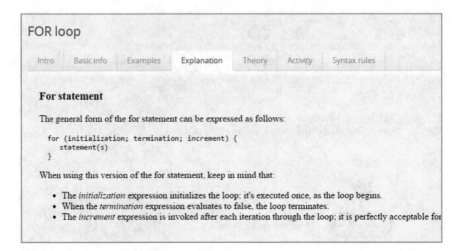

Fig. 9.17 Lesson for verbal learners

Sequential learners prefer to go through the course gradually, in a linear way with each step followed logically from the previous one, while *Global* learners tend to learn in large leaps, sometimes skipping learning objects and jumping to material that is more complex.

According to these characteristics of learning styles, *Sequential* learners go through lessons by in advance predefined order (Klašnja-Milićević et al. 2011) while the *Global* learners are provided with the possibility to freely jump through the courseware. To define order of concepts, next rules are implemented (rules are defined only for the adaptation to a sequential learning style because learners with a global learning style are presented learning material in a predefined sequence):

`Learner(?x) ∧ hasLearningStyle(?x,explanation) →` `hasPrerequisite(loopStatements,explanation)`	AR20
`Learner(?x) ∧ hasLearningStyle(?x,sequential) →` `hasPrerequisite(executionControl,loopStatements)`	AR21
`Learner(?x) ∧ hasLearningStyle(?x,sequential) →` `hasPrerequisite(classes,executionControl)`	AR22
`itd.`	

Adaptation rules AR20, AR21 and AR22 are used to define presentation layout for sequential learners. Where *hasPrerequisite* is a data valued property atom that defines prerequisites among resources. Based on the defined prerequisites, system can make decision whether to present one lesson in time in sequential order (for *Sequential* learners) or to present links to all lessons at ones to learner (in case of *Global* learners).

In the first case, the interface elements for sequential navigation (in our case the buttons for *Next/Previous resource/lesson*) will be shown to a learner (Fig. 9.18).

Fig. 9.18 Navigation for sequential learners

Fig. 9.19 Elements for
global learners

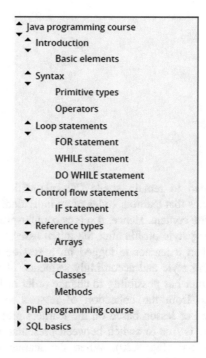

On the other hand, interface elements for non-sequential navigation will be presented (Fig. 9.19) for *Global* learner.

Adaptation rules, which are used to form the learner model and update *Learner model ontology* will be presented in next section.

9.3.3   Rules for Building Learner Model

This section describes several examples of methods that can be used for a learning style modelling. The adaptive feedback in existing e-learning systems is usually based solely on an initial assessment of the learning style profile, which is then

Fig. 9.20 Experience bar

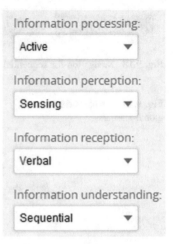

Information processing:

Active ▼

Information perception:

Sensing ▼

Information reception:

Verbal ▼

Information understanding:

Sequential ▼

expected to remain stable (Klašnja-Milićević et al. 2011). However, research indicates that learning styles of an individual can vary depending on the task or the learning content. Hence, it seems counter-productive to lock the learner into a fixed learning style profile after the initial assessment

When a learner is logged in, a session is initiated based on learner specific learning style and accordingly sequence of lessons are recommended to him/her. A learner has possibility to change order of lessons (s)he attends. After selecting a lesson, from the collection of lessons available, system chooses presentation method of lesson based on the learners preferred style. For the rest of the lesson, learner is free to switch between presentation methods by using the media experience bar (Fig. 9.20). When the learner completes the sequence of learning materials, the system evaluates his/her knowledge degree for each lesson.

Following rule updates learner model (Vesin et al. 2011):

Learner(?x) ∧ Interaction(?y) ∧ hasInteraction(?x,?y) ∧ Resource(?r) ∧ resourceUsed(?y,?r) ∧ Performance(?p) ∧ hasResult(?y,?p) ∧ hasGrade(?p,?m) ∧ swrlb:greaterThan(?m, 1) → isLearned(?r, true) ∧ hasPerformance(?x,?p)	AR23

With the rule AR23, system is using recorded results of learner's interaction, earned grade and data about used resources to memorize learner's performance in the session. Variables *x, y, r, m* and *p* present *Learner, Interaction, Resource, Grade* and *Performance*.

Resource presents a learning object which has been accessed by the learner in the current session.

Meaning of the rule AR23 is: if the learner interacts with specific concept and during that interaction (s)he took the test and earned specific grade, than system should memorize that learner's performance. In addition, *isLearned* property of that particular concept should be set to true.

Rule AR23 is only executed when learner earn positive grade.

If learner shows insufficient knowledge, next rule is executed:

`Learner(?x) ∧ Interaction(?y) ∧ hasInteraction(?x,?y) ∧`
`Resource(?r) ∧ resourceUsed(?y,?r) ∧ Performance(?p) ∧`
`hasResult(?y,?p) ∧hasGrade(?p,?m) ∧ swrlb:equal(?m, 1) →`
`hasExecuted(?x,?r) ∧ hasPerformance(?x,?p)`
	AR24

Previous rule marks concept as executed but learned status is still left negative, meaning that new concept that supports same learning object will be used in next iteration.

If learner does not provide required level of performance results within session with presentation method used for certain learning style category, his/her current learning style category will be modified by next rule:

`Learner(?x) ∧ hasLearningStyle(?x,verbal) ∧ Interaction(?i)`
`∧ hasInteraction(?x,?i) ∧ Resource(?r) ∧ resourceUsed(?i,?r)`
`∧ ResourceRole(?s) ∧ hasRole(?r,?s) ∧ supports(?s, verbal) ∧`
`Performance(?p) ∧ hasResult(?i,?p) ∧ hasGrade(?p, grade) ∧`
`swrlb:lessThan(grade, required) → hasLearningStyle(?x,`
`visual)`
	AR25

Variables *x, i, r, s* and *p* present *Learner, Interaction, Resource, Resource role* and *Performance*, respectively. Meaning of the rule is: if in any time of the execution of system, exists learner with *Verbal* learning style which interacts with system and during that interaction (s)he had accessed appropriate resource but not earned sufficient grade (required grade level is kept in global value *required*), than, learning style of that learner should be changed.

If initial learning style for learner was visual, than next rule would be executed:

`Learner(?x) ∧ hasLearningStyle(?x, visual) ∧ Interaction(?i)`
`∧ hasInteraction(?x,?i) ∧ Resource(?r) ∧ resourceUsed(?i,?r)`
`∧ ResourceRole(?s) ∧ hasRole(?r,?s) ∧ supports(?s, visual) ∧`
`Performance(?p) ∧ hasResult(?i,?p) ∧ hasGrade(?p, grade) ∧`
`swrlb:lessThan(grade, required) → hasLearningStyle(?x,`
`verbal)`
	AR26

Rule AR26 launches for learners with *visual* learning style. The meaning of the rules is: if learner *x* with *Verbal* learning style interacts with system and during that interaction (s)he had accessed appropriate resource *r* but not earned sufficient grade (required grade level is kept in global value *required*), than, learning style of that learner should be changed to other style from *Information reception* domain: i.e. *Visual* learning style. That implies that in next session, learner will be presented with resources that are defined to support that new learning style category.

Similar rules will be executed for other categories of learning styles (intuitive/sensing, global/sequential and active/reflective).

9.3.4 Adaptation Based on Resource Sequencing

Resource sequencing is a well-established technology in the field of intelligent tutoring systems (Janssen et al. 2007). The idea of resource sequencing is to generate a personalized course for each learner by dynamically selecting the most optimal teaching actions, presentation, examples, task or problems at any given moment. By optimal teaching action it is considered an operation that in the context of other available operations brings the learner closest to the ultimate learning goal. Most often the goal is to learn and acquire some knowledge up to a specific level in an optimal amount of time. However, it is easy to imagine other learning goals, such as minimizing learner error rates in problem solving.

Adaptation rules that generate recommendations in general tutoring system model can be divided into two categories:

- off-line rules,
- recommendation rules.

The details of the recommendation process and examples of implemented rules are presented in the rest of the section.

9.3.4.1 Off-Line Rules

Off-line rules use data from *Learner model ontology* on-the-fly to recognize learners' goals and content profiles. Learners are grouped into clusters, i.e. individual learning styles categories within the four defined domains of learning styles. Based on the results of the questionnaires it was possible to define 16 (2^4) clusters. Clusters are used to identify common features and activities of learners from the same cluster (Vesin et al. 2011).

At the beginning of the learning process in our system, Protus 2.1 distinguishes cluster that learner belongs to with one of the appropriate rules. System generates a list of recommended learning materials and activities to learners from the same cluster.

For example, learners from cluster *cl1* belong to following categories: *active*, *sensitive*, *visual* and *sequential* within the domains: *Information processing*, *Information perception*, *Information reception* and *Information understanding*, respectively. Then, learners from cluster *cl2* belong to following categories: active,

Table 9.3 Learning styles clusters

Clusters	Cluster 1	Cluster 2	Cluster 3	Cluster 4	...	Cluster 16
Information processing	Active	Active	Active	Active		Reflective
Information perception	Sensing	Sensing	Sensing	Sensing		Intuitive
Information reception	Visual	Visual	Verbal	Verbal		Verbal
Information understanding	Sequential	Global	Sequential	Global		Global
Cluster's code	cl1	cl2	cl3	cl4	...	cl16

sensitive, visual and global learners within the domains: *Information processing*, *Information perception*, *Information reception* and *Information understanding*, respectively. Table 9.3 contains a complete list of the clusters and corresponding learning styles.

Corresponding rule is defined for each cluster. Example of rule that is triggered if learner belongs to cluster cl1 (rule AR27) and cluster cl2 (rule AR28) are (analog rules are used for other clusters):

Learner(?x) ∧ Performance(?p) ∧ hasPerformance(?x,?p) ∧ Condition(?c) ∧ generates(?p,?c) ∧ BehaviourPattern(?b) ∧ include(?p,?b) ∧ isTypeOf(?b,?n) ∧ NavigationSequence(?n) ∧ consistsOf(?n,a1) ∧ consistsOf(?n,b2) ∧ consistsOf(?n,c1) ∧ consistsOf(?n,d1) ∧ consistsOf(?n,e4)∧ swrlb:greatherThen (grade, required) → belong(?x,?cl1)	AR27
Learner(?x) ∧ Performance(?p) ∧ hasPerformance(?x,?p) ∧ Condition(?c) ∧ generates(?p,?c) ∧ BehaviourPattern(?b) ∧ include(?p,?b) ∧ isTypeOf(?b,?n) ∧ NavigationSequence(?n) ∧ consistsOf(?n,a1) ∧ consistsOf(?n,b2) ∧ consistsOf(?n, d1) ∧ consistsOf(?n, e4) ∧ consistsOf(?n, c1)∧ swrlb:greatherThen (grade, required) → belong(?x,?cl2)	AR28

Rules AR27 and AR28 determine that learner belongs to the appropriate pattern, based on navigation sequence *n* that contains learning resources {a1, b2, c1, d1, e4} and {a1, b2, d1, e4, c1}, respectively. This sequence is a list that contains: taken tutorial, visited examples, tasks and tests taken, etc. In this case, determined patterns are *cl1* and *cl2*, respectively.

Variables *x*, *p*, *c*, *b* and *n* present *Learner*, *Performance*, *Condition*, *BehaviorPattern*, and *Navigation Sequence*, respectively.

Condition class collects data about learner's performance and his/her learning style and generates appropriate type of performed personalization that will be implemented. Generated personalization, in fact, presents specific navigational pattern recommended to learner.

Meaning of the rule is: if in any time of the execution of the system, exists learner which interacts with system under specific condition and during that interaction (s)he successfully completed navigational sequence (predefined for specific behaviour pattern), than that sequence can be treated as appropriate for that learner and (s)he should be put in adequate cluster. Pattern discovering is only executed if learner successfully completes navigation sequence, that is to say, if learner has earned sufficient grade.

During course, system updates the database of learners' interaction with the system. This database contains interaction data to build *sequential patterns*.

Next rule is used to update the navigation pattern:

`Learner(?x) ∧ Concept(?c) ∧ Resource(?r) ∧ hasResource(?c,?`
`r) ∧ isLearned(?c, true) ∧ hasPerformance(?x,?p) ∧`
`BehaviourPattern(?b) ∧ NavigationSequence(?n) ∧ isTypeOf(?`
`b,?n) → swrl:add(?n,?r,?p)`
	AR29

Adaptation rule AR29 adds visited resource to navigation pattern for the current session.

SWRL function `swrl:add(?x,?y,?z)` adds resource *y* and details about learner *z* to navigation pattern *x*.

Meaning of rule AR29 is: if at any point of the systems' execution exists a learner who has successfully mastered a concept that contains particular resource, then system should add that resource and details about learner to the successful navigation pattern.

9.3.4.2 Recommendation Rules

Recommendation rules produce a list of recommended learning objects. From the existing list of learning content and based on the discovered sequences of educational resources, the list of recommended actions and recourses is sent to alter learner-system interaction within a new session.

The recommendation module is design to create a recommendation list according to the ratings of these frequent sequences, provided by the system (Vesin et al. 2012). Patterns are ranked based on assessment of learners after visit to resources in specific order.

For example, if the learner is determined to belong to cluster *cl1*, it means that (s) he is attended sequence of resources: {a1, b2, c1, d1, e4}.

Based on that initial sequence, Protus 2.1 rated highest extended set of resources: {a1, b2, c1, d1, e4, f2} and, therefore, resource *f2* is recommended to him/her.

Recommendations are generated with following adaptation rules:

`Learner(?x) ∧ Performance(?p) ∧ hasPerformance(?x,?p) ∧`
`Condition(?c) ∧ generates(?p,?c) ∧ BehaviourPattern(?b) ∧`
`include(?p,?b) ∧ isTypeOf(?b,?n) ∧ NavigationSequence(?n) ∧`
`belong(?x,?cl1) → isRecommended(f2,true)`
	AR30

Meaning of the rule AR30 is: if in any time of the execution of the system, exists learner whom specific navigation sequence of resources has been recommended (with specific condition) than system should recommend to him/her next specific resource that belongs to that navigational sequence. Recommendation status of the resource *f2* is set to true, therefore link to that resource is annotated or highlighted. Wether student is following the recommended path or not, does not influence the rules itself, but influence execution of rules in next sessions because ratings of frequent navigation sequences are calculated after every session.

Ratings of frequent sequences are not calculated only by sequences followed by a student itself but earned grades throughout session are also included in calculation. Therefore, every system-imposed path still counts towards placing the learner in a particular cluster.

The above SWRL rules can be executed using the Jess rules engine after providing the factual knowledge. The system uses the Jess's Java API that allows the creation of *Jess rule engine* instances. An instance of this class loads all defined rules and check the terms defined in them and starts execution of rules if all conditions are met. After firing the rule, the inferred knowledge can be written back to the ontology repository and update the knowledge base (Chi 2009). Rules will automatically start immediately after all requirements defined in them are fulfilled. After firing the rule, the inferred knowledge can be written back to the ontology repository and update the knowledge base. Whereas ontologies were used to increase interoperability and reusability of domain information, rules were employed to represent the adaptation logic in a way that teachers can inspect, understand and modify the rationales behind adaptive functionalities.

9.4 Course Development

Model and architecture of general tutoring system described in this book consists of clearly defined adaptable, expandable and separated components (Vesin et al. 2012). System enables easy modification of adaptation and personalization of learning materials that are offered to learners. Formally defined ontologies will allow the reuse of tutoring system components for implementation of similar systems.

General tutoring system model enables the development of courses in different domains in three phases (Fig. 9.21):

- creation of skeleton application with use of Vaadin Java framework (Grönroos 2010),
- creation of individual courses, appropriate teaching materials for each course as well as a set of appropriate tests for assessment of acquired learners' knowledge,
- presentation of personalized learning materials to each individual learner.

Three phases of course development provides a clear separation of the activities of three groups of system designers:

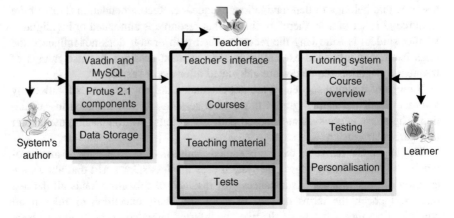

Fig. 9.21 The development of courses from different domains in general tutoring system architecture

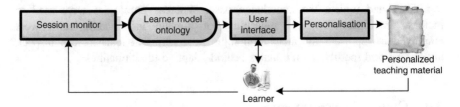

Fig. 9.22 Learners' activities in general tutoring system model

- designers of the system's architecture and user interface
- authors who take care of personalization that is performed in tutoring system and
- authors of learning materials.

Complete separation of the content and its design on one hand and application design and presentation of materials on the other hand, allows the development of courses in different domains and easy reuse of individual components of application. Different courses follow the same learning process that contains: monitoring learners' activities, development of appropriate learner model and personalization of learning materials (Fig. 9.22).

Defined general tutoring system architecture enables the development of courses from different domains supported with various forms of user interface personalisation and adaptation of educational material. Next chapter presents concretisation of the general tutoring system architecture for generating a Java programming course within the system Protus 2.1.

References

Aroyo, L., & Dicheva, D. (2004). The new challenges for e-learning: The educational semantic web. *Educational Technology and Society*.

Aroyo, L., & Mizoguchi, R. (2003). Process-aware authoring of web-based educational systems. In *CAiSE Workshops*.

Bechhofer, S. (2009). OWL: Web ontology language. In *Encyclopedia of database systems* (pp. 2008–2009). Berlin: Springer.

Borland, J. (2007). A smarter web. *Technology Review, 110*(2), 64–71. http://doi.org/Article

Brusilovsky, P. (2004). KnowledgeTree: A distributed architecture for adaptive e-learning. In *WWW Alt. '04: Proceedings of the 13th International World Wide Web Conference on Alternate Track Papers and Posters* (pp. 104–113). http://doi.org/10.1145/1013367.1013386

Carmagnola, F., Cena, F., Gena, C., & Torre, I. (2005). A semantic framework for adaptive web-based systems. In *SWAP* (Vol. 166, pp. 82–97).

Chi, Y.-L. (2009). Ontology-based curriculum content sequencing system with semantic rules. *Expert Systems with Applications, 36*(4), 7838–7847.

Devedzic, V. (2006). *Semantic web and education* (Vol. 11). http://doi.org/10.1007/978-0-387-35417-0

Felder, R., & Silverman, L. (1988). Learning and teaching styles in engineering education. *Engineering Education, 78*, 674–681. http://doi.org/10.1109/FIE.2008.4720326

Fensel, D. (2004). *Ontologies: A silver bullet for knowledge management and electronic-commerce* (p. 162). Berlin: Spring.

Grönroos, M. (2010). *Book of Vaadin: Vaadin 6.4. Writing*.

Horrocks, I., Patel-Schneider, P. F., Boley, H., Tabet, S., Grosof, B., Dean, M., et al. (2004). SWRL: A semantic web rule language combining OWL and RuleML. *W3C Member Submission, 21*, 79.

Janssen, J., den Berg, B., Tattersall, C., Hummel, H., & Koper, R. (2007). Navigational support in lifelong learning: enhancing effectiveness through indirect social navigation. *Interactive Learning Environments, 15*(2), 127–136.

Klašnja-Milićević, A., Vesin, B., Ivanovic, M., & Budimac, Z. (2011). Integration of recommendations and adaptive hypermedia into java tutoring system. *Computer Science and Information Systems, 8*(1), 211–224. http://doi.org/10.2298/CSIS090608021K

Kolb, D. (1984). Individuality in learning and the concept of learning styles (pp. 61–98). Englewood Cliffs, New Jersey: Prentice Hall.

Popescu, E., Bădică, C., & Trigano, P. (2007). Rules for learner modeling and adaptation provisioning in an educational hypermedia system. In *Proceedings—9th International Symposium on Symbolic and Numeric Algorithms for Scientific Computing, SYNASC 2007* (pp. 492–499). http://doi.org/10.1109/SYNASC.2007.72

Romero, C., Ventura, S., Hervas, C., & Gonzalez, P. (2006). Rule mining with {GBGP} to improve web-based adaptive educational systems. In *Data mining in e-learning* (Vol. 4, pp. 171–188). Retrieved from http://library.witpress.com/pages/listPapers.asp?q_bid=392

Soloman, B. A., & Felder, R. M. (2005). Index of learning styles questionnaire. *NC State University*. Available Online at: http://www.Engr.Ncsu.Edu/learningstyles/ilsweb.Html. Last Visited on May 14, 2010.

Strohmaier, M., Walk, S., Pöschko, J., Lamprecht, D., Tudorache, T., Nyulas, C., ... Noy, N. F. (2013). How ontologies are made: Studying the hidden social dynamics behind collaborative ontology engineering projects. *Web Semantics: Science, Services and Agents on the World Wide Web, 20*, 18–34.

Ullrich, C. (2004). Description of an instructional ontology and its application in web services for education. In *Proceedings of Workshop on Applications of Semantic Web Technologies for E-learning, SW-EL* (Vol. 4, pp. 17–23).

Vesin, B., Ivanovic, M., Klašnja-Milićević, A., & Budimac, Z. (2011). Rule-based reasoning for altering pattern navigation in programming tutoring system. In *2011 15th International Conference on System Theory, Control, and Computing (ICSTCC)* (pp. 1–6).

Vesin, B., Ivanović, M., Klašnja-Milićević, A., & Budimac, Z. (2012). Protus 2.0: Ontology-based semantic recommendation in programming tutoring system. *Expert Systems with Applications, 39*, 12229–12246. http://doi.org/10.1016/j.eswa.2012.04.052

Vesin, B., Ivanović, M., Klašnja-Milićević, A., & Budimac, Z. (2013). Ontology-based architecture with recommendation strategy in Java tutoring system. *Computer Science and Information Systems, 10*(1), 237–261. http://doi.org/10.2298/CSIS111231001V

Wand, Y., Storey, V. C., & Weber, R. (1999). An ontological analysis of the relationship construct in conceptual modeling. *ACM Transactions on Database Systems (TODS), 24*(4), 494–528.

Part IV
Case Study: Design and Implementation of Programming Tutoring System

Chapter 10
Design, Architecture and Interface of Protus 2.1 System

Abstract General tutoring system model, presented in previous chapter, can be used as a skeleton for an implementation of concrete programming tutoring system. This chapter presents details about implementation of Java programing course based on defined model. Protus 2.1 is a tutoring system designed to provide learners with personalized courses from various domains. It is an interactive system that allows learners to use teaching material prepared for appropriate courses and also includes parts for testing acquired knowledge. In spite of the fact that this system is designed and implemented as a general tutoring system, the first completely implemented and tested version was for an introductory Java programming course. This chapter presents the most important requests for implementation of personalization options in e-learning environments, as well as design, architecture and interface of Protus 2.1 system. Details about previous versions of the system, defined user requirements for the new version of the system, architecture details, as well as general principles for application of defined general tutoring model for implementation of programming course in Protus 2.1 are presented.

General tutoring system model, presented in Chap. 9 presents a skeleton for an implementation concrete programming tutoring system. This section presents details about implementation of Java programing course based on defined model.

Protus 2.1 is a tutoring system designed to provide learners with personalized courses from various domains. It is an interactive system that allows learners to use teaching material prepared for appropriate courses and also includes parts for testing acquired knowledge. In spite of the fact that this system is designed and implemented as a general tutoring system, the first completely implemented and tested version of the system was for an introductory Java programming course (Vesin et al. 2009). Java is chosen because it is a clear example of an object-oriented language and it is therefore suitable for the teaching of the concepts of object-orientation. The course is designed for learning programming basics for learners with no previous object-oriented programming experience.

Protus 2.1 fulfills three primary goals, identified by earlier exploration in this field (Jones et al. 2006). The first goal was to provide a personalized tutoring

© Springer International Publishing Switzerland 2017
A. Klašnja-Milićević et al., *E-Learning Systems*,
Intelligent Systems Reference Library 112, DOI 10.1007/978-3-319-41163-7_10

system for learners in a platform independent manner. The second goal was to provide the teachers with useful reports identifying the strengths and weaknesses of the learner's learning process. Finally, the third goal was to provide a rapid development tool for creating basic elements of tutoring system: new learning objects, units, tutorials and tests.

General tutoring system model, defined in Chap. 9, enables implementation of an unlimited number of personalized courses from different domains and defining formal rules for adaptation of educational materials to each individual learner.

This chapter will summarize the general setup of Protus 2.1 system before discussing the recommendation module in detail in Chap. 11. After reviewing and illustrating current state of the art in this area in Sect. 10.1. Section 10.2 presents previous versions of Protus system, while programming course in Protus 2.1 is described in the Sect. 10.3. Section 10.4 describes development of ontologies specifically for a programming course.

10.1   Personalised Programming Tutoring Systems

Computer technology has been used to develop a wide range of educational software, from early computer-based training systems to Web-based adaptive hypermedia, multimedia courseware, and educational games. These systems have given learners access to a great variety of pedagogical approaches that supplement classroom learning and provide items outside the classroom. This variety has been helpful in reaching learners who don't do well with traditional lecture and textbook instruction. Our attention was focused only on a specific kind of tutoring systems. In the rest of this section, we first describe programming tutoring systems in general, then we present tutoring systems that use different recommendation techniques in order to suggest the most appropriate online learning activities for learners, based on their preferences, learning style, knowledge and the browsing history of other learners with similar characteristics.

10.1.1   Programming Tutoring Systems

Most of the tutoring systems for learning programming languages found on the Web are more or less only well-reformatted versions of lecture notes or textbooks (Vesin et al. 2009). As a consequence, in these systems are not implemented interactivity and adaptivity.

The functions that such systems can perform vary. Some of them are used for learner assessment like Java Bugs (Suarez and Sison 2008) and JITS (Sykes and Franek 2003; Sykes 2007). Also, some of them are adaptive Web-based tutorials (García et al. 2009). One step further in implementation of adaptation was made by systems like JOSH-online (Bieg and Diehl 2004), iWeaver (Wolf 2003) and

CIMEL ITS (Blank et al. 2005; Wei et al. 2005). JavaBugs examines a complete Java program and identifies the most similar correct program to the learner's solution among a collection of correct solutions. After that it builds trees of misconceptions using similarity measures and background knowledge (Suarez and Sison 2008). The developers of this system focused on the construction of a bug library for novice Java programmer errors, which is a collection of commonly occurring errors and misconceptions.

The WWW-based introductory LISP course ELM-ART (ELM Adaptive Remote Tutor) is based on ELM-PE (Brusilovsky et al. 1996), an on-site intelligent learning environment that supports example-based programming, intelligent analysis of problem solutions, and advanced testing and debugging facilities. For annotating the links, the authors use the traffic light metaphor. A red ball indicates pages which contain information for which the user lacks some knowledge, a green ball indicates suggested links, etc. Java Intelligent Tutoring System—JITS is a tutoring system designed for learning Java programming (Sykes and Franek 2003). JITS implements Java Error Correction Algorithm (JECA), an algorithm for a compiler that enables error correction intelligently changing code, and identifies errors more clearly than other compilers. This practical compiler intelligently learns and corrects errors in learners' program (Sykes 2007). iWeaver is an interactive Web-based adaptive learning environment, developed as a multidisciplinary research project at RMIT University Melbourne, Australia (Wolf 2003). iWeaver was designed to provide an environment for the learner by implementing adaptive hypermedia techniques to teach the Java programming language. It implements several established adaptation techniques, including link sorting, link hiding and conditional page content. The current version of iWeaver does not support adaptive navigation, which is one of the best researched areas of adaptive environments. JOSH is an interpreter for the Java programming language (Bieg and Diehl 2004) originally designed to make easier teaching Java to beginners. Recently the interpreter was restructured into a server based interpreter applet and integrated into an online tutorial on Java programming called JOSH-online. CIMEL ITS is an intelligent tutoring system that provides one-on-one tutoring to help beginners in learning object-oriented analysis and design. It uses elements of UML before implementing any code (Blank et al. 2005). A three-layered Learner model is included which supports adaptive tutoring by deducing the problem-specific knowledge state from learner solutions, the historical knowledge state of the learner and cognitive reasons about why the learner makes an error (Wei et al. 2005). This Learner model provides an accurate profile of a learner so that the intelligent tutoring system can support adaptive tutoring.

Most of the existing e-learning platforms for teaching programming have not yet taken the advantage of adaptivity (Emurian 2006; Holland et al. 2009; Sykes and Franek 2003), possibly because the expected profit has not justified the high effort of implementing and authoring adaptive courses. Moreover, most of the adaptive tutoring systems do not support e-learning standards. Our system recommends a media experience that is most likely to be chosen in the current learning context by the current learner. This recommendation mechanism is attempting to accommodate a possible variation in a learner's learning style profile. Also, up to now most, if not all systems do not take into consideration the important aspect of learning style preferences or how and when to adjust the presented topic based on the preferred presentation method of the learner.

10.1.2   Tutoring Systems with Implemented Recommendation

A personalized recommender system that uses Web mining techniques for recommending a learner, which (next) links to visit within an adaptive educational hypermedia system was described in (Romero et al. 2006). They presented a specific mining tool and a recommender engine that they have integrated in the AHA! system, in order to help the teacher to carry out the whole Web mining process. They made several experiments with real data in order to show the suitability of using together, clustering and sequence mining algorithms, for discovering personalized recommendation links (Brusilovsky et al. 1996).

Another system described in (Soonthornphisaj et al. 2006) allows all learners to collaborate and exchange their expertise in order to predict the most suitable learning materials to each learner. This smart e-learning system applies the collaborative filtering approach (Soonthornphisaj et al. 2006) that has an ability to predict the most suitable documents to the learner. All learners have the chance to introduce new material by uploading the documents to the server or pointing out the Web link from the Internet and rate the currently available materials. My Online Teacher 2.0 (MOT 2.0) successfully combines Web 2.0 features (such as tags, rating system, feedback, etc.) in order to support both learners and teachers in personalized systems (Ghali and Cristea 2009). Ghali and Cristea focus on a study which can explain how to use and combine more effectively the recommendation of peers and content adaptation to enhance the learning outcome in e-learning systems.

In the last few years, some research studies have been conducted on developing an approach that identifies learning styles from learners' behavior in an online course (Arenas-García et al. 2007; García et al. 2007; Sabine Graf et al. 2010). The rationale is that adapting courses to the learning preferences of the learners has a positive effect on the learning process, leading to an increased efficiency, effectiveness and/or learner satisfaction (Popescu 2010). The adaptive response of existing environments is often restricted to pictures and text instead of multimedia presentations, with some exceptions like the iWeaver (Wolf 2003). Systems like Logic-ITA, ProGuide and Jeliot 3 gave us good ideas and perspective which functionalities could be included in new Web-based tutoring system (Merceron and Yacef 2004; Myller 2006). Compared to current tutoring systems which only execute on a standalone machine (JavaBugs, JITS, CIMEL ITS, Jeliot 3) or have just basic interactivity and adaptivity implemented (JOSH-online, Java Bugs, Logic-ITA), Protus 2.1 system integrates content and link adaptation in order to accomplish completely functional Web-based tutoring system with personalization capability. Protus 2.1, as an e-learning system, offers a constant on-the-fly adaptation of the course units and their presentation to the current needs and preferences of the individual learner. This guarantees a significant, individual success of a learner. None of the above mentioned systems implement full use of the recommender techniques (like collaborative filtering, association rule mining and clustering), just the basic data mining techniques. Second, besides learning content ranking and tagging, Protus 2.1 also supports learning path generation and personalization based on the learning style identification. Our work differs from previous mentioned papers in several aspects. First, we combine several adaptation techniques, both recommendations of material and adaptive hypermedia, in order to personalize the lessons presentation to learners. Second, besides learning content ranking and tagging, Protus 2.1 also supports learning material clustering and learning path generation. Third, despite the great variety of tutoring systems in the literature we chose to focus our attention on a programming tutoring system that defines scalable and adaptable architecture. Protus 2.1 provides the possibility to import knowledge from various domains so that the process of learning can be performed in whatever domain of knowledge. This choice enabled us to develop a system for knowledge presentation and acquisition that tries to be independent of the specific domain.

10.2 Previous Versions of Protus 2.1

Protus 2.1 is improved and enhanced version of its predecessors Mag and Protus Systems (Table 10.1):

- **Mag tutoring system**, which is used for presentation of the basic concepts of the Java programming language as well as to assess the knowledge of learners.

Table 10.1 Functionalities of previous versions

System's functionalities	Mag	Protus	Protus 2.1
Java online course	x	x	x
Reports on the progress of learners	x	x	x
Communication between learners and mentors	x	x	x
Adding new learning materials	x	x	x
Possibility to enter additional courses		x	x
Integrated systems for generating recommendations		x	x
Learning style identification		x	x
Reuse and sharing of learning material		x	x
Semantic Web technologies integrated			x
Possibilities for adding new adaptation methods			x
Tag-based recommendation			x

Mag system did not include any of recommender systems for the implementation of the course personalization (Vesin et al. 2009).

- **Protus system** (abbreviation of: Programming Tutoring System) supported recommendations of learning materials based on the identified learning styles of each individual learner. Protus was designed as general online tutor for learning content from different domains.

10.2.1 Mag System

Mag is a tutoring system designed to help learners in learning programming languages (Klašnja-Milićević et al. 2009). The first completely proposed and tested version was for an introductory Java programming course. The main idea was that a learner who was attending a course become familiar with the Java programming language basics, its syntax, basic elements and commands. It is an interactive system that allows learners to use teaching material prepared for programming languages within courses and to test their knowledge. The system offers multiple options: a review of course materials, testing of knowledge, online programming, etc.

Mag system is based on a centralized architecture, which includes two clearly separate parts of the system (Fig. 10.1):

- server-side application with a database that includes lessons, examples, tests and data on each individual learner,
- series of Web pages that form an interactive course.

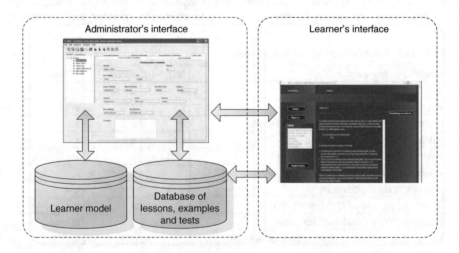

Fig. 10.1 Mag system architecture

The server side of the system contains application for system administrators (learner's mentors) (Fig. 10.2). The basic functionalities of this application are:

- review and update of learners' database (adding and deleting data on learners)
- reports on the learners' progress (passed lessons, taken tests, success in particular lessons, etc.),
- review and update database of administrators (mentors)
- receiving reports on the progress of learners (learners are divided into classes, each class is assigned by a mentor),
- search options on learners' data with different criteria,
- communication with learners (review of received and sent messages),
- adding new units, tutorials, examples and tests designed for specific lessons.

The system uses and updates the database of learners, administrators, tests and lessons. Messages, sent between learners and mentors, as well as tutorials, lessons, examples and tests are kept in the appropriate file system.

Online course consists of a series of .jsp pages intended for a learner, which allow logging, signing of new learners, presentation of education material, testing knowledge, review the of learners' success, and communication with the mentor (Fig. 10.3).

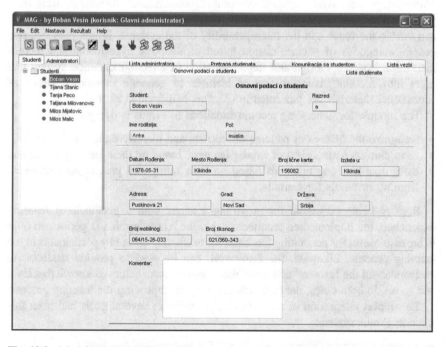

Fig. 10.2 Administrator's interface of Mag

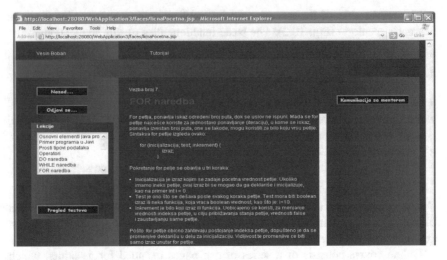

Fig. 10.3 Mag system

10.2.2 Protus System

Protus is designed as an extension of existing Web-based Java tutoring system—
Mag. Components which support different recommendation techniques were integrated in Protus in order to allow standardization and formalization of the content
and enable the reuse and the interoperability of the systems. Protus provides two
general categories of personalization based on adaptive hypermedia and recommendation systems. Design and implementation of a recommendation system that
takes into account characteristics of learners to generate recommendations of
educational materials are presented in (Klašnja-Milićević et al. 2011).

The module for generating recommendations in Protus is designed to:

- recognize the behaviour patterns of learners and identify their learning styles,
- form clusters (categories) of similar learners, based on their learning styles and
- categorize teaching materials based on their rating and present recommended
  learning materials for learners.

Besides being beneficial for providing learners with a personalized learning
experience, the implemented architecture and the reasoning that is performed over
it, are also useful for generating feedback for teachers—other key participants in the
learning process. Likewise, the framework can be used to provide feedback to
teachers about the learners' activities, their performance, achieved knowledge level
and so on. In both cases, the feedback can help in improving the learning process.

To support integration of recommendation technics several goals had been fulfilled in Protus system:

- separation of two different interfaces—for learners and teachers,

- a strict separation of different modules: domain, application, adaptation and learner model, in order to ensure a good modularization of the system components,
- continuous monitoring of learner progress and development of a dynamic learner model,
- enabling communication and collaboration among learners and between learners and teachers,
- assessment of knowledge and increasing competency level of learners,
- functionalities for creation of new learning content and migration of content from external sources,
- semantically rich descriptions of the components' functionality, in order to allow effective interoperability among system components, and
- providing effective coordination and communication between the system components.

10.2.2.1   System's Architecture of Protus

The architecture of Protus is based on experiences gained from similar Web-based learning systems (Chen et al. 2010; Merino and Kloos 2008; Šimić 2004) and architecture for ontology-supported adaptive Web-based education systems suggested in De Bra et al. (2003), Devedzic (2006). Figure 10.4 presents the general architecture of redesigned and extended Protus system. Protus system consists of five functional modules: *domain module*, *learner model*, *application module*, *adaptation module* and *session monitor*.

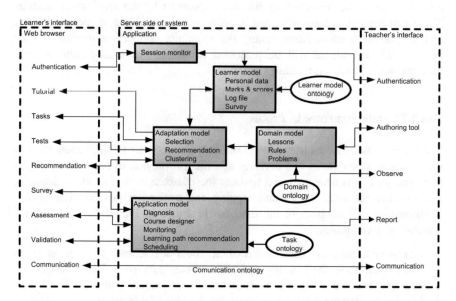

Fig. 10.4 Protus 2.1 system architecture

The *domain module* presents the storage for all essential learning material, tutorials and tests. It describes how the information content is structured.

The *adaptation module* is responsible for building and updating learner's model characteristics and also for personalization of content presented to the learner. It processes changes of learner's characteristics based on learner's activities and it provides an adaptation of the visible aspects of the system for specific learner. Its main tasks also include storage and management of learning material, presenting that material to learners, generating of reports and test results etc.

Each *learner model* is a collection of both static and dynamic data about the learner. Static data include personal data, specific course objectives, etc. Dynamic data include scores, time spent on specific lesson, marks, etc. Above mentioned data, the learner model contains also a representation of the learner's performance and learning history. The system uses that information in order to predict the learner's behaviour, and thereby adapt to his/her individual needs.

Within *session monitor* component, the system gradually re-builds the learner model during the session, in order to keep track of the learner's actions and his/her progress, to detect and correct his/her errors and possible to redirect the session accordingly. At the end of the session, all of learners' preferences are recorded in learner model. The learner may change this information at any time by editing his/her preferred learning style. Therefore, if a learner does not agree with the system's assumptions about his/her preferences, (s)he can inspect his/her learner model and make changes in it during learning sessions. The learner model is then used along with other information and knowledge to initialize the next session for the same learner.

The *application module* performs the adaptation. To be exact, the adaptation module follows the instructional directions specified by the application module. These two components are separated in order to make adding new content clusters and adaptation functionalities easier. For example, application module creates decision which material will be presented to learner while adaptation module presents chosen material to learn.

10.2.2.2   Data Structure in Protus

The architecture of Protus uses a Java DataBase Connectivity (JDBC) connection to the database which stores and retrieves specific information of concert to the system. The Protus system uses and updates the database with data about learners, teachers, course, unit, the lessons, tagging and evaluation process. The Enitity Relationship—ER diagram of the database is shown in Fig. 10.5. The database consists of seven tables:

1. **Learner**. It contains basic information about the learner as well as some information about the learning styles and learner's progress (Table 10.2).
2. **Teacher**. It contains basic information about the teacher (Table 10.3).
3. **Lesson**. It contains information about the lessons (Table 10.4).

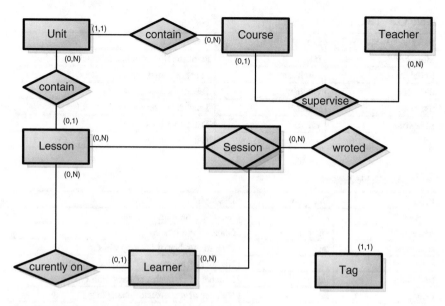

Fig. 10.5 ER diagram of Protus system database

Table 10.2 Table learner

Field	Type	Description
idlearner	Int	Identification number of learners
name	varchar	Learner's name
surname	varchar	Learner's surname
username	varchar	Learner's username
password	varchar	Learner's password
gender	varchar	Gender of learner
year	int	Year of birth
adr	varchar	Learner's adress
processing	int	Category within 'Information Processing' domain
perception	int	Category within 'Information Perception' domain
reception	int	Category within 'Information Reception' domain
understanding	int	Category within 'Information Understanding' domain
begin_time	date	Date of the course beginning
overall_time	time	Total duration of the course
avg_grade	decimal	Average grade of learner
percentage	decimal	Percentage of course completed
lesson	int	Number of completed lesson
curlesson	int	Current lesson

Table 10.3 Table teacher

Field	Datatype	Description
idteacher	Int	Identification number of teacher
firstname	varchar	Teacher's first name
lastname	varchar	Teacher's last name
username	varchar	Teacher's username
password	varchar	Teacher's password
title	varchar	Teacher's title

Table 10.4 Table *Lesson*

Field	Datatype	Description
idlesson	int	Identification number of lesson
name	varchar	Name of the lesson
resources	int	Overall number of learning objects
intro	int	Number of resources of introduction type
basic info	int	Number of resources of basic info type
example	int	Number of resources of example type
explanation	int	Number of resources of explanation type
theory	int	Number of resources of theory type
activity	int	Number of resources of activity type
syntax	int	Number of resources of syntax type
unit	int	Identification number of unit

4. **Unit**. It contains information about the unit, lesson and learning objects (resources) from which lesson is consisted (Table 10.5).
5. **Course**. It contains information about the course, units, lessons, the number of learners attending the course and duration of the course (Table 10.6).

Table 10.5 Table *Session*

Field	Datatype	Description
idsession	int	Identification number of session
learner	int	Identification number of learner who has completed session
lesson	int	Lesson visited during session
sessiontime	time	Duration of the session
grade	int	Earned grade

Table 10.6 Table *Unit*

Field	Datatype	Description
idunit	int	Identification number of unit
name	varchar	Name of the unit
course	int	Identification number of course
lesson	int	Number of lessons

6. **Session**. It contains information about learner sessions that the learner has completed during the course and the grades (s)he earned for them (Table 10.7).
7. **Tag**. It contains information about tags and information about lessons and learning objects for which the tag is placed (Table 10.8).

10.2.2.3 User Interface in Protus

Two main roles exist in the system, intended for two types of system's users:

- **Learners**—they are attending the course and use the system in order to gain certain knowledge and
- **Teachers and content authors**—the lessons and learner database administrators. They track learning process of learners and help them with their assignments, as it will be presented later in the section.

Therefore, as in previous version of the system—in Mag, separated user interfaces are provided for learners and teachers (Vesin et al. 2009). Teacher's interface helps in process of managing data about a learner and course material. This component has not undergone major changes over the previous version of the system. Improvements have been made only in terms of advanced features reporting on the progress of learners.

Table 10.7 Table *Course*

Field	Datatype	Description
idcourse	int	Identification number of course
name	varchar	Name of the course
unit	int	Number of units
lesson	int	Number of lessons
LSsuported	int	Learning styles which are suported
learner_num	int	Number of learners attending the course
duration	varchar	Duration of the course
teacher	int	Id of supervisor

Table 10.8 Table *Tag*

Field	Datatype	Description
idtag	int	Identification number of tag
idlearner	int	Identification number of learner
lesson	int	Identification number of lesson
Learning_object	int	Identification number of learning object
value	varchar	Entered tag
session	int	Identification number of session

Learner's interface is a series of Web pages that provide two options: taking lessons and testing learner's knowledge. All data about learner and his progress in the course, as well as data about tutorials, tests and examples are stored in the system's server.

10.2.2.4   Learner's User Interface in Protus

For every lesson the same sequence of activities has to be followed. At the beginning of a lesson, participants are shown a short introductory text on the lesson's topic (Fig. 10.6), additionally explained with appropriate examples (Fig. 10.7). At the end of each lesson a post-test is conducted. Test contains several multiple-choice questions and Protus provides feedback on their answers and gives the correct solutions after every submitted answer.

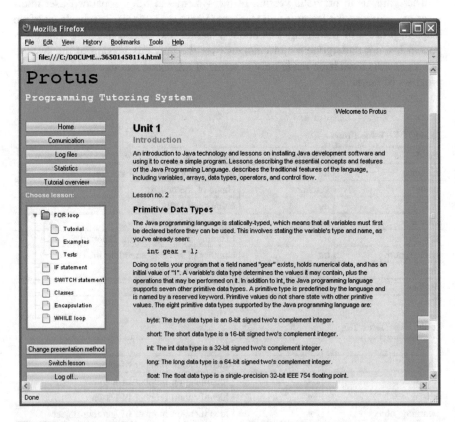

Fig. 10.6 Lessons tutorial

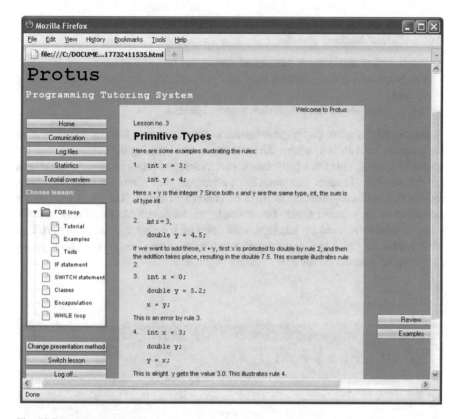

Fig. 10.7 Lessons example

10.2.2.5   Teacher's User Interface in Protus

Besides being beneficial for providing learners with personalized learning experience, Protus system is also useful for generating feedback for other participants in the learning process-content authors and teachers. Content authors are typically subject matter experts who create learning content, that is subsequently used by teachers who wrap that content into a learning design. Protus can be used to provide feedback to teachers about the learners' activities, their performance, achieved collaboration level and the similar activities. In both cases, the feedback helps in improving the learning process.

Protus aims at helping teachers rethink the quality of the learning content and learning design of the course they teach. To this end, the system provides teachers with feedback about the relevant aspects of the learning process taking place in the online learning environment they use. The provided feedback is based on the analyses of the context data collected in the learning environment. In particular, Protus informs teachers about:

- the activities the learners performed during the learning process,
- the usage of learning content they had prepared and deployed in the tutoring system,
- the peculiarities of the interaction among learners.

Figure 10.8 depicts the graphic interface that represents the real interaction of the teachers with Protus, where the assessment and tracking data and statistics are generally shown. This form facilitates data retrieval and provides appropriate results for the teacher. Teacher can combine parameters and filters in order to obtain reports that will be presented in form of charts and tables. The chart type varies according to the selected filters. For example, the teacher could know what specific material was more used by learners, what kind of learning style they preferred or what grades they earned for every particular lesson. These reports can show results for group of learners or for every learner separately.

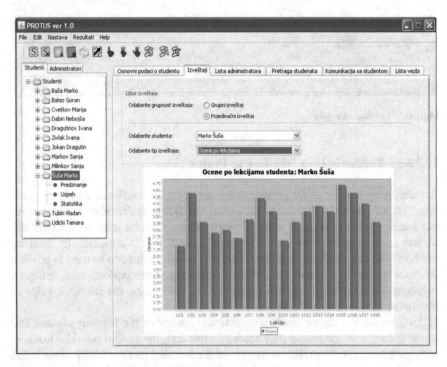

Fig. 10.8 Teachers' interface in Protus

10.3   Protus 2.1

The general tutoring system model presented in Chap. 9 presented the basis for development of new and improved version of Protus. It is important to note that the original architecture of Mag and Protus did not bring any kind of homogenous representation of components. Each one was represented by different formats, using a variety of tools. Thus, for example, the format of learning materials in Protus had to be changed in order to optimize the system, so it was not possible to use some of the existing material. Also, the use of various technologies prevents reuse and sharing of educational material with other systems. To make the system more widely available and to allow its easier development and upgrading, the need arose to represent each component of the system in form of the ontology. Each component will be responsible for specific tasks. This approach will make it easier to understand the role of each component and, consequently, to promote interoperability among the components of the architecture. The developed system is highly modular, which allows better flexibility and future replacement of various components as long as they comply with the current interface.

System Protus 2.1 enables the development of courses from different domains. Each course consists of a series of lessons (concepts) showing the individual segments of the domain being processed. Each lesson consists of a series of resources that represent files with descriptions of individual parts of lessons (introduction, explanations, examples, tasks, exercises, etc.). Each lesson is linked to one or more appropriate tests to check learners' knowledge. Based on the results of tests level of learner progress is determined, learner model is updated and further personalization options are generated.

The main aim of Protus 2.1 was to improve adaptation of the teaching material according to demand and need of each individual learner. This version of the system supports personalization options in the form of recommendation systems. Also, for the first time, Vaadin framework was used to develop system's components. Vaadin is an open source Web application framework for rich Internet applications. In contrast to JavaScript libraries and browser-plugin based solutions, it features a server-side architecture, which means that the majority of the logic runs on the servers (Grönroos 2010).

Protus 2.1, like previous versions of this system, consists of two basic components: user interface for learners and user interface for teachers.

10.3.1   Learner's Interface

Learners attend courses through the Web interface implemented in Protus 2.1 system (Fig. 10.9). The user interface of this system offers the learner the following functionalities:

- review of the offered courses and teaching materials,

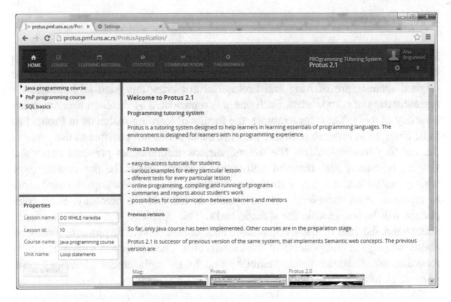

Fig. 10.9 User interface of Protus 2.1

- various display formats of teaching materials adapted to different learning styles,
- testing of acquired knowledge,
- communication with the mentor and other learners,
- reports about progress, test results, coursework and their own learning styles.

Organization of pages within Protus 2.1 system is presented in Fig. 10.10. Sequence of pages changes depending on whether the learner first entered the system or continuing his/her course. Details of applied personalization and use of Protus 2.1 will be presented in Chap. 11.

10.3.2   User Interface for Teachers and Course Administrators

Besides being beneficial for providing learners with personalized learning experience, Protus 2.1 system is also useful for generating feedback for other participants in the learning process—administrator and/or teachers (Klašnja-Milićević et al. 2011). Protus 2.1 can be used to provide feedback to teachers about the learners' activities, their performance, achieved collaboration level and the similar activities. In both cases, the feedback helps in improving the learning process. Teachers have access to special functions within Protus 2.1 system. There are two levels of privileges. First is the higher level—a level with unlimited possibilities. A teacher

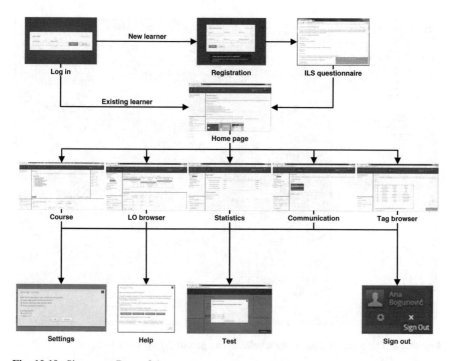

Fig. 10.10   Site map—Protus 2.1

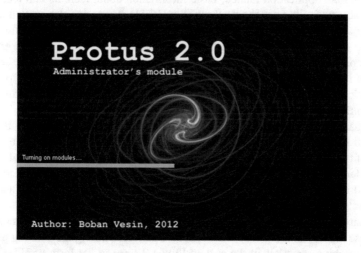

Fig. 10.11   Application's intro screen

with this privilege level can enter information about new teachers and data necessary for their connection. Other teacher's activities are limited to editing the
learners' data. All teachers can access and modify the data of all active learners.

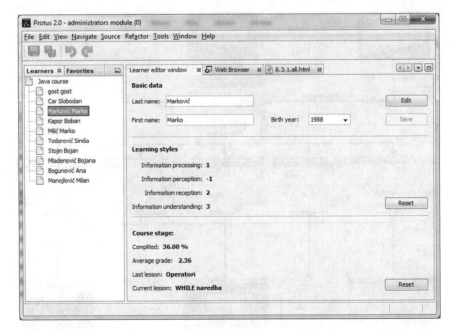

Fig. 10.12 The main window of the administrative part of the system

During the application launch, basic information about learners and teachers is loaded automatically from the database (Fig. 10.11). These data are entered into the appropriate lists. Protus 2.1 aims at helping teachers rethink the quality of the learning content and learning design of the course they teach. To this end, the system provides teachers with feedback about the relevant aspects of the learning process taking place in the online learning environment they use. The provided feedback is based on the analyses of the context data collected in the learning environment.

In panel *Learners* list of all learners can be loaded from the database (Fig. 10.12). *Edit* button opens a dialog for editing information about the learner. Each of the fields is ready to enter the new data. Data on gender, class, and country of residence of the learner is entered by selecting one of the options. If an unauthorized user wants to make changes to the data, the system prevents it and (s)he receives a warning message.

In particular, Protus 2.1 provides statistical reports to the teacher. Teachers can use these reports to inform about the learner's activities during the learning process.

All teaching materials in the system Protus 2.1 are in the form of HTML documents. The system loads the document specified in that moment for active learners on the basis of data from the model learner and displays it in a Web browser.

Administrator's module of the system Protus 2.1 has integrated HTML editor that allows the creation and review of teaching materials (Fig. 10.13).

Lessons generated with *html* editor can be previewed within integrated Web browser in Protus 2.1 administrative module (Fig. 10.14).

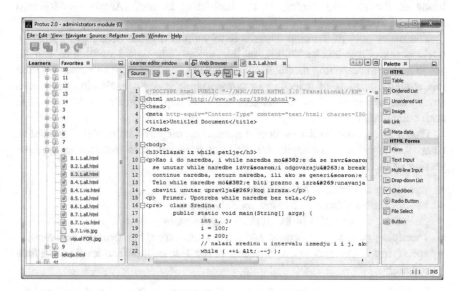

Fig. 10.13 Entry of new HTML resource

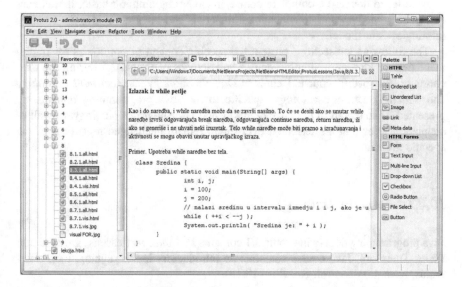

Fig. 10.14 Overview of teaching material

10.4   Development of Ontologies for Java Programming Course

Authoring of adaptation and personalization is actually authoring of learner models and applying different adaptation strategies and techniques to ensure efficient tailoring of the learning content to the individual learners (Aroyo and Riichiro Mizoguchi 2003).

The major goal of learning systems is to support a given pedagogical strategy (Dehors and Faron-Zucker 2006). In this scope pedagogical ontologies can be associated with reasoning mechanisms and rules to enforce a given strategy. Often this strategy consists of selecting or computing a specific navigation sequences among the resources. Thus, formal semantics are required in Protus 2.1 to enable such computation.

Protus 2.1 is built on the basis of a general tutoring system model presented in Chap. 9. Educational ontologies for different purposes were included in the new version of the system, such as for:

- presenting a domain—*domain ontology*,
- building learner model—*learner model ontology*,
- presenting of activities in the system—*task ontology*,
- specifying pedagogical actions and behaviours—*teaching strategy ontology*,
- specifying behaviours and techniques at the learner interface level—*interface ontology*.

System's ontologies in Protus 2.1 are written in OWL. Protégé tool was used for development of ontologies and their translation into OWL (Protégé 2011).

In order to develop a complete course in Java programming basic, it was necessary to create appropriate educational materials and integrate it into the existing tutoring system model.

The following subsections present the specific content of ontologies for implementation of Java programming course and activity of Protus 2.1 during the execution of this specific course. Details of *Domain ontology*, *Learner model ontology* and *Teaching strategy ontology*, specific to the Java programming domain are presented. Data stored in the *Task ontology* and *User interface ontology,* as well as adaptation rules that are executed over them, remain unchanged regardless of the specific domain.

10.4.1   Domain Ontology

Java programming course in Protus 2.1 contains 18 *Concepts* (lessons) grouped into *Units*. Therefore, Java course contains: an introductory lesson, syntax, loop statements, execution control, etc. (Fig. 10.15).

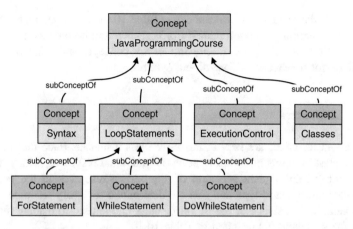

Fig. 10.15 An excerpt of ontology as domain topology of Protus 2.1

Each concept can be assigned any number of different resources (text files, images, animations, etc.). All resources are assigned depending on their Resource type. So we have: theory, examples, assignments, exercises, syntax rules, and so on. Resources in Protus 2.1 are stored in the form of *HTML* documents.

10.4.1.1   Concepts

An excerpt of a domain ontology covering basics of Java programming concepts with *subConceptOf* relationships between these concepts has been shown In Fig. 9. 14 (Vesin et al. 2013). This figure depicts the root concept with some of its sub concepts: *Syntax*, *LoopStatements*, *ExecutionControl* and *Classes*. The *LoopStatements* concept is further specialized and fine-grained into *ForStatement*, *WhileStatement* and *DoWhileStatement*. Clear specification of other relations between concepts will be useful for further personalization purposes.

New lessons in domain ontology are created by a new instance of the *Concept* class. An example of *Concept* class instance that is used to collect information about the *ForStatement* concept is presented in Table 10.9.

Table 10.9 Example of instance of *Concept* class *Concept*

Property description	Property name	Property value	Property type
Concept's id	hasId	C009	Datatype property
Concept's name	hasName	For Statement	Datatype property
Resource's type	hasResource	R017	Object property
Superclass	subConceptOf	loopStatements	Object property
Prerequisite	hasPrerequisite	ExecutionControl	Object property
Prerequisite	hasPrerequisite	Syntax	Object property

This particular instance of *Concept* class (Table 10.9) has unique id: C009. It has been used for defining a lesson named *ForStatement* and it contains data about its superclass (it is *subConcept* of *loopStatements* concept) and concepts that are prerequisite for it (*ExecutionControl* and *Syntax*).

10.4.1.2   Resources

Details about resources are kept in *Resource* class instances. Each instance of the *Resource* class contains basic information on individual resources, which will later be used for the subsequent selection of appropriate resources in the process of personalization. Specific type and role are determined for every resource.

An example of instance of the *Resource* class that is used to display the syntax rules of *for* statement is presented in Table 10.10.

This particular instance of *Resource* class has unique id: R017. It is used for presenting a syntax rule for a lesson (concept) named *ForStatement* and it contains a link to a certain *jpg* file (Fig. 10.16) that will be presented to the learner if the system chooses this resource during personalization activities. All resources are grouped by their type, role and the concept they support and these groups present a basis for successful recommendation during the personalization process.

10.4.2   Learner Model Ontology

During a learning session, the learner interacts with a tutoring system. Learner interactions can be used to draw conclusions about his/her possible interests, goals, tasks, knowledge, etc. These conclusions can be used later for providing personalization. Ontology for learner observations should therefore provide a structure of information about possible learner interaction.

Table 10.10 Example of instance of the *Resource* class

Property description	Property name	Property value	Property type
Resource's id	hasId	R017	Datatype property
Resource's name	hasName	forLoop017	Datatype property
Resource's type	isTypeOf	Syntax rule	Object property
Concept's type	isResourceFor	For Statement	Object property
Resource's role	supports	Visual style	Datatype property
Is resource visited?	isVisited	yes	Datatype property
Is resource recommended	isRecommended	no	Datatype property
File Type	hasFileType	jpg	Datatype property
Concept's role	hasRole	definition	Datatype property
Link to used figure	hasFigure	Figure 6.jpg	Annotation properties

Fig. 10.16 Figure resource
in Protus 2.1

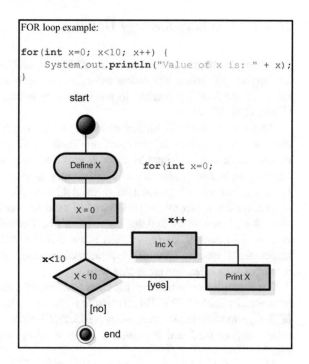

```
FOR loop example:

for(int x=0; x<10; x++) {
    System.out.println("Value of x is: " + x);
}
```

All actions of learners in Protus 2.1 are recorded in *Interaction* class. The
interaction involves all relevant actions of the learner during particular session
(*Session* class).

An example of instance of *Interaction* class that is used to keep track of the
learner's interaction during one session is presented in Table 10.11.

This particular instance of *Interaction* class has unique id: I007. It is formed
during session S1012 when learner with id L01 took test and gained results, which
are all collected in instance of *Performance* class with id P01. Instances of
Performance class contain, among other, data about grade that learner earned
during current testing. Based on that *Performance* data, system makes decision of
further personalization within *Teaching strategy ontology* described in Sect. 10.4.3
(adaptation rules: AR23 do AR26).

Table 10.11 Example of instance of *Interaction* class

Property description	Property name	Property value	Property type
Interaction's id	hasId	I007	Datatype property
Session	partOf	S1012	Object property
Interaction's type	hasType	T02 (test)	Object property
Used concept	conceptUsed	C032	Object property
Learner	whoInteracted	L01	Object property
Performance	hasResult	P01	Object property

10.4.3   Teaching Strategy Ontology

Authoring of adaptation and personalization is actually authoring of learner models and applying different adaptation strategies and techniques to ensure efficient tailoring of the learning content to the individual learners and their learning styles (Vesin et al. 2012).

An instance of the *Condition* class that was formed based on the *Performance* data of every learner is presented in Table 10.12. This particular instance of *Condition* class has unique id I006. It contains data collected based on learner's learning style and performance. The instance is populated with adaptation rules presented in Sect. 9.3 (adaptation rules: AR1 i AR2).

The system monitors the movement of each learner through the learning materials. Adaptation rules updates instance of the *BehaviourPattern* in the manner presented in Sect. 9.3.3 (adaptation rules: AR29 i AR30). Instance of this class represents a specific type of *NavigationalSequence* class that contains array of resources which learner interacted with.

For example, an instance of the class *BehaviourPattern* contains the information presented in Table 10.13. This particular instance of *BehaviourPattern* class is type of *NavigationalSequence* marked as NS02, that learner made during session S1012, with rating of 0.37 and it generates instance of *Personalization* class marked as PR09.

The presence of these specific instances of *Personalisation* classes means that the currently active learner will be presented with teaching material intended for learners with visual learning style.

Table 10.12 Example of instance of *Condition* class

Property description	Property name	Property value	Property type
Condition's id	hasId	I006	Datatype property
LS of learner	generatedBy	LS03	Object property
LS Category	hasLearningStyleCategory	visual	Datatype property
LS domain	hasLearningStyleDomain	Information Reception	Datatype property
Learner's performance	generatedBy	P01	Object property
Personalization	Generates	PR09	Object property

Table 10.13 Example of instance of *BehaviourPattern* class

Property description	Property name	Property value	Property type
Behaviour pattern's id	hasId	BP0016	Datatype property
Navigational sequence	isTypeOf	NS02	Object property
Session	partOf	S1012	Datatype property
Personalization	generate	PR09	Object property
Ratings of navigational sequence	hasRate	0,37	Datatype property

All personalization activities within Protus 2.1 are performed based on previously mentioned data in instances of *Condition* and *BehaviourPattern* classes. Chapter 11 will present a few examples of performed personalization based on all collected data about learner's interaction with the system (adaptation rules: AR27 i AR28).

10.4.4   Task Ontology and User Interface Ontology

Task ontology defines roles of certain concepts and resources while U*ser interface ontology* creates an array of resources that are recommended and presented to active learner. Therefore, the data stored in these ontologies and adaptation rules executed over them, remain unchanged regardless of the specific domain.

References

Arenas-García, J., Meng, A., Petersen, K. B., Lehn-Schioler, T., Hansen, L. K., & Larsen, J. (2007). Unveiling music structure via plsa similarity fusion. In *2007 IEEE Workshop on Machine Learning for Signal Processing* (pp. 419–424).

Aroyo, L., & Riichiro, M. (2003). Authoring support framework for intelligent educational systems. In *AIED-2003* (pp. 362–364).

Bieg, C., & Diehl, S. (2004). Educational and technical design of a web-based interactive tutorial on programming in Java. *Science of Computer Programming, 53*(1), 25–36.

Blank, G., Parvez, S., Wei, F., & Moritz, S. (2005). A web-based ITS for OO design. In *Workshop on Adaptive Systems for Web-Based Education Tools and Reusability* (Vol. 12).

Bra, P. de, Aerts, A., Berden, B., Lange, B. de, Rousseau, B., Santic, T., et al. (2003). AHA! The adaptive hypermedia architecture. In *Proceedings of the Fourteenth ACM Conference on Hypertext and Hypermedia—HYPERTEXT '03, 4*, 81. doi:10.1145/900065.900068

Brusilovsky, P., Schwarz, E., & Weber, G. (1996). ELM-ART: an intelligent tutoring system on World Wide Web. In *Intelligent Tutoring Systems* (pp. 261–269).

Chen, J. M., Chen, M. C., & Sun, Y. S. (2010). A novel approach for enhancing student reading comprehension and assisting teacher assessment of literacy. *Computers and Education, 55,* 1367–1382. doi:10.1016/j.compedu.2010.06.011

Dehors, S., & Faron-Zucker, C. (2006). Qbls: a semantic web based learning system. In *World Conference on Educational Multimedia, Hypermedia and Telecommunications (ED-MEDIA).*

Devedzic, V. (2006). *Semantic web and education. Book* (Vol. 11). http://doi.org/10.1007/978-0-387-35417-0

Emurian, H. H. (2006). A web-based tutor for Java (TM): evidence of meaningful learning. *International Journal of Distance Education Technologies, 4*(2), 10.

García, P., Amandi, A., Schiaffino, S., & Campo, M. (2007). Evaluating Bayesian networks' precision for detecting students' learning styles. *Computers and Education, 49*(3), 794–808.

García, E., Romero, C., Ventura, S., & De Castro, C. (2009). An architecture for making recommendations to courseware authors using association rule mining and collaborative filtering. *User Modeling and User-Adapted Interaction, 19*(1–2), 99–132.

Ghali, F., & Cristea, A. I. (2009). MOT 2.0: a case study on the usefulness of social modeling for personalized e-learning systems. In *AIED* (pp. 333–340).

Graf, S., & Ives, C., et al. (2010). A flexible mechanism for providing adaptivity based on learning styles in learning management systems. In *2010 IEEE 10th International Conference on Advanced Learning Technologies (ICALT)*, (pp. 30–34).

Grönroos, M. (2010). *Book of Vaadin: Vaadin 6.4. Writing*.

Holland, J., Mitrovic, A., & Martin, B. (2009). J-LATTE: a constraint-based tutor for java.

Jones, N., Macasek, M., Walonoski, J., Rasmussen, K., & Heffernan, N. (2006). Common tutor object platform—An e-learning software development strategy. In *Proceedings of the 15th international conference on World Wide Web, Edinburgh, Scotland* (pp. 307–316).

Klašnja-Milićević, A., Vesin, B., Ivanovic, M., & Budimac, Z. (2011). Integration of recommendations and adaptive hypermedia into java tutoring system. *Computer Science and Information Systems, 8*(1), 211–224. doi:10.2298/CSIS090608021K

Klašnja-Milićević, A., Vesin, B., Ivanović, M., & Budimac, Z. (2009). Integration of recommendations into Java tutoring system. In *The 4th International Conference on Information Technology ICIT 2009 Jordan*.

Merceron, A., & Yacef, K. (2004). Mining student data captured from a web-based tutoring tool: initial exploration and results. *Journal of Interactive Learning Research, 15*(4), 319.

Merino, P. J. M., & Kloos, C. D. (2008). An architecture for combining semantic web techniques with intelligent tutoring systems. In *Intelligent Tutoring Systems* (pp. 540–550).

Myller, N. (2006). Automatic prediction question generation during program visualization. In *Proceedings of the Fourth Program Visualization Workshop*.

Popescu, E. (2010). Adaptation provisioning with respect to learning styles in a w-based educational system: an experimental study. *Journal of Computer Assisted Learning, 26*(4), 243–257.

Protégé. (2011). *The Protégé Ontology Editor. Financial Executive (Vol. 19)*. http://doi.org/10.5121/ijait.2011.1401

Romero, C., Ventura, S., Hervas, C., & Gonzalez, P. (2006). Rule mining with {GBGP} to improve web-based adaptive educational systems. In *Data mining in e-learning* (Vol. 4, pp. 171–188). Retrieved from http://library.witpress.com/pages/listPapers.asp?q_bid=392

Šimić, G. (2004). The multi-courses tutoring system design. *Computer Science and Information Systems, 1*(1), 141–155.

Soonthornphisaj, N., Rojsattarat, E., & Yim-Ngam, S. (2006). Smart e-learning using recommender system. In *Computational Intelligence* (pp. 518–523). Springer.

Suarez, M., & Sison, R. (2008). Automatic construction of a bug library for object-oriented novice java programmer errors. In *Intelligent Tutoring Systems* (pp. 184–193).

Sykes, E. (2007). Developmental process model for the Java intelligent tutoring system. *Journal of Interactive Learning Research, 18*(3), 399.

Sykes, E. R., & Franek, F. (2003). An intelligent tutoring system prototype for learning to program java TM.

Vesin, B., Ivanović, M., & Budimac, Z. (2009). Learning management system for programming in java. *Annales Universitatis Scientiarum De Rolando Eötvös Nominatae, Sectio-Computatorica, 31*, 75–92.

Vesin, B., Ivanović, M., Klašnja-Milićević, A., & Budimac, Z. (2012). Protus 2.0: ontology-based semantic recommendation in programming tutoring system. *Expert Systems with Applications, 39*, 12229–12246. doi:10.1016/j.eswa.2012.04.052

Vesin, B., Ivanović, M., Klašnja-Milićević, A., & Budimac, Z. (2013). Ontology-based architecture with recommendation strategy in Java tutoring system. *Computer Science and Information Systems, 10*(1), 237–261. doi:10.2298/CSIS111231001V

Wei, F., Moritz, S. H., Parvez, S. M., & Blank, G. D. (2005). A student model for object-oriented design and programming. *Journal of Computing Sciences in Colleges, 20*(5), 260–273.

Wolf, C. (2003). iWeaver: towards' learning style'-based e-learning in computer science education. In *Proceedings of the Fifth Australasian Conference on Computing Education-Volume 20* (pp. 273–279).

Chapter 11
Personalization in Protus 2.1 System

Abstract The ultimate goal of developing Protus 2.1 system has been increasing the learning opportunities, challenges and efficiency. Two important ways of increasing the quality of Protus 2.1 service are to make it intelligent and adaptive. Different techniques need to be implemented to adapt content delivery to individual learners according to their learning characteristics, preferences, styles, and goals. Protus 2.1 provides two general categories of personalization in system based on adaptive hypermedia and recommender systems: content adaptation and adaptation of user interface. Several approaches are used to personalize the material presented to the learner. Programming course in Protus 2.1 offers three types of personalization to each individual learner: (1) use of recommender systems, (2) learning styles personalization and (3) personalization based on resource sequencing. This chapter presents Protus 2.1 functionalities as well as personalization options from the end-user perspective.

The ultimate goal of developing Protus 2.1 system has been increasing the learning opportunities, challenges and efficiency. Two important ways of increasing the quality of Protus 2.1 service are to make it intelligent and adaptive.

Different techniques need to be implemented to adapt content delivery to individual learners according to their learning characteristics, preferences, styles, and goals. Protus 2.1 provides two general categories of personalization in system based on adaptive hypermedia and recommender systems: (Ivanović et al. 2008):

- *Content adaptation*—presenting the content in different ways, according to the domain module and information from the learner model. All learners and contents are grouped into classes of similar objects in order to recommend optimum resources and pathways. The principle of clustering is maximizing the similarity inside an object group and minimizing the similarity between the object groups.

  Such clusters needed to be defined in Protus 2.1 in order to provide learner with the most suitable learning material and to form the most suitable pathway. The system maintains different versions of pages it presents to the learners or in some cases different versions of page fragments within the page, it selects the adequate version to be presented to the learner according to the information in learner model.

System also hides advanced content from a novice learner and shows suitable additional content to more advanced learner. Personalisation is based on the fact that learners with different learning styles will master the content easier if it is presented to them in an appropriate way, for example with block diagrams or graphic representation of the syntax rules rather than with textual descriptions.

- *Adaptation of user interface* involves adaptation of certain user interface elements displayed on Web pages in order to present recommendations of certain teaching materials.

  User interface adaptation in Protus 2.1 is performed in the form of *link adaptation*. The system modifies the appearance and/or availability of every link that appears on a course Web page, in order to show the learner whether the link leads to interesting new information, to new information the learner is not ready for, or to a page that provides no new knowledge. System makes some links inaccessible to the learner if the system estimates from the learner model that such links take him/her to the irrelevant information.

  System assumes that less successful learners will be interested in additional materials. Therefore, those learners may click the link for additional material on the interface.

Several approaches can be used to personalize the material presented to the learner. Programming course in Protus 2.1 offers three types of personalization to each individual learner:

- use of recommender systems (Klašnja-Milićević et al. 2011a)
- learning styles personalization (Klašnja-Milićević et al. 2011b) and
- personalization based on resource sequencing (Vesin et al. 2012).

Application of those three techniques will be explained in subsequent sections.

11.1 The Protus 2.1 Component for Making Recommendations

A recommender System in e-learning environments assists learners in discovering relevant learning actions and educational material that perfectly match their profile, at the right time, in the right context, and in the right way, keep them motivated and enable them to complete their learning activities in an effective and efficient way (Tang and McCalla 2005).

The Protus 2.1 component for building automatic recommendations is composed of three modules (Fig. 11.1):

- *A learner-system interaction module*, which pre-processes data to build learner models. The information from learners' registration form and learning style

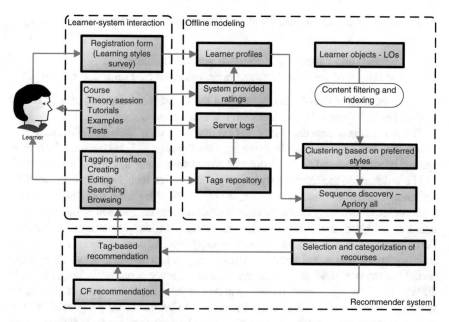

Fig. 11.1 The recommendation component

survey are collected in order to create an initial personal profile. The data about all learners' activities like sequential patterns, visited pages (tutorials, theory session or examples), test results and grades earned are collected within this module and saved into the server logs. The functionality available by clicking on an active learning object includes searching and categorization, as well as the ability to add tags or notes, and to modify/delete selected tags or notes. All information about tagging process are kept in tags repository and used for tag-based recommendation.

- *An off-line module*, which uses learner models on-the-fly to recognize learners' goals and content profiles. After appropriate learning style is determined for each learner, based on the initial survey, learning content is filtered, depending on the current status of the course, learner's affiliation and learners' tags. The offline module is launched periodically to perform all necessary calculations (user/cluster assignment, evaluation computation). Calculations are performed using a current snapshot of input data, and all results are stored until all procedures are finished. The previously computed results are left intact at this stage. Upon completion, the old results are overwritten with the newly calculated data. This allows the recommendation engine, to be very quick, with computational complexity kept at minimum. That is because all requirements to be done online are to look up the user in the users/clusters assignment database and retrieve ranks and recommendations for that given cluster.

- *A recommendation engine*, which produces a recommendation list according to the:

  - learners' and experts' tags for each generated cluster and
  - the ratings of the frequent sequences, provided by Protus 2.1 system

From the filtered list of learning content the list of recommended actions is sent to alter learner–system interaction within a new session.

When learners access the system for the first time, their learning styles need to be tested. The Felder-Silverman learning style model (FSLSM) is considered the most appropriate to be used in a computer-based educational system (Kinshuk et al. 2011). It describes the learning style on a more detailed level than the other models. By using dimensions instead of types, the strengths of learners' preference towards a particular learning style can be represented. Moreover, FSLSM is based on tendencies, enabling the learning style model to consider exceptional behaviour. Furthermore, FSLSM is widely used in adaptive learning systems focusing on learning styles and some researchers even argue that it is the most appropriate model for the use in adaptive learning systems (Carver et al. 1999; Kuljis and Liu 2005). Based on this model a corresponding psychometric assessment instrument was created. It was called the Felder-Solomon's Index of Learning Styles (ILS). It is a 44-item questionnaire where learners' personal preferences for each dimension are expressed with values between +11 and −11 per dimension, with steps ±2. This range comes from the eleven questions that are posed for each dimension (Graf 2007). This style indicates a preference for some presentation methods over others.

According to the different combinations of learning styles it is possible to define clusters, which determined learner profiles. These results are stored in the learner model, which are used for the adaptation in Protus 2.1 (Klašnja-Milićević et al. 2011b).

When a learner is registered or logged in, (s)he can begin the process of learning. A learning session is initiated based on the learner's specific learning style and sequence of lessons is recommended to him/her. The learner can change the order of lessons (s)he is attending. After selecting a lesson, from the collection available in Protus 2.1, the system chooses a presentation method for the lesson based on the preferred style. For the rest of the lesson, learners are free to switch among presentation methods using the media experience bar, which will be explained in detail in Sect. 11.2. When the learner completes the sequence of learning contents, the system evaluates the learner's knowledge degree. The test contains several multiple-choice questions and code completion tasks. Protus 2.1 then provides feedback to the learner on his/her answers and gives the correct solutions after the test. Recommendations cannot be made for the entire set of (all) learners in the same way, because even for learners with similar learning interests, their ability to solve a task can vary due to variations in their knowledge level. In our approach, we perform a data clustering technique as a first step to cluster learners based on their learning styles. These clusters are used to identify coherent choices in frequent sequences of learning activities.

Recommendation list can be created according to the ratings of these frequent sequences, provided by Protus 2.1 system. Also, a recommendation list can be created according to the learners' and experts' tags in every cluster, separately. During the learning process learner can tag each learning object. The details of the whole personalization process are presented in the rest of the Chapter.

11.2 Learning Style Identification in Protus 2.1

It is obvious that different learners have different preferences, needs and approaches to learning. Psychologists call these differences individual learning styles. Therefore, it is very important to take into account these different learning styles when implementing learning environments in order to make educational process more efficient. Learning styles can be defined as unique manners in which learners begin to concentrate on, process, absorb, and retain new and difficult information (Dunn et al. 1984; Pritchard 2013).

The term learning styles refers to the concept that individuals differ in regard to what mode of instruction or study is most effective for them (Pashler et al. 2009). Proponents of learning-style assessment argue that optimal instruction requires diagnosing individuals' learning style and tailoring instruction accordingly. While many learning style models exist in literature, in Protus 2.1 system we implemented model by (Felder and Silverman 1988) concerning the different Cognitive Styles of Learning (CSL) learners may have, which were described in (Klašnja-Milićević et al. 2011b). Based on those CSL, a GUI interface was developed to enable the learner himself/herself to categorize his/her CSL, set his/her learning goals and characteristics of work environment and the kind of course (s)he wants to take. At run-time, the learner model is updated taking into account the learning activities.

Before initial session, and after learning style has been determined by the ILS, current learning style category of the particular learner must be written in learner model of Protus 2.1. The learning style will be further investigated (and updated if necessary) by observing a pattern in the choices (s)he makes.

11.2.1 Adaptation Process in Protus 2.1

There are over seventy identifiable approaches to investigate and/or describe learning style preferences. As we already mentioned, we used one such data collection instrument, called *Index of Learning Styles—ILS* (Felder et al. 2000). The ILS is a 44 question, freely available, multiple-choice learning styles instrument, which assesses variations in individual learning style preferences across four dimensions or domains. These are *Information Processing, Information Perception,*

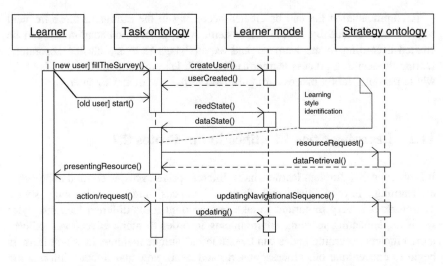

Fig. 11.2 Adaptation based on the learning styles

Information Reception, and *Information Understanding.* Within each of the four domains of the ILS there are two categories of learners:

- **Information Processing**: *Active* and *Reflective* learners,
- **Information Perception**: *Sensing* and *Intuitive* learners,
- **Information Reception**: *Visual* and *Verbal* learners,
- **Information Understanding**: *Sequential* and *Global* learners.

In the rest of the section adaptation process in Protus 2.1 based on learning styles identification is presented.

At the beginning of the learning session Protus 2.1 requests information about the status of the course from the *Learner model ontology* for the particular learner (Fig. 11.2). This data includes information about the current lesson and the learning style category of the learner within one of the four domains of the ILS. Request for appropriate resources which will be presented to the learner, based on this data, is sent to the *Teaching strategy ontology.* Further, all activities of learners are monitored, as well as all requests (s)he send to the system and a *Learner model ontology* is updated accordingly.

If the learner does not provide the required level of performance results within a session based on the presentation method used for his/her certain learning style category, his/her current learning style category will be modified. In those cases, the system changes the current learning style of the learner to its alternative, from the same domain. Learning styles are grouped in pairs (active and reflective, sensing and intuitive, visual and verbal, sequential and global), therefore every learning style has only one alternative within the domain. The established threshold of knowledge can change over time depending on the level of learners' knowledge.

For example, if a learner with *Verbal* learning style interacts with the system and during that interaction (s)he had accessed appropriate concept, but not earned sufficient grade (required grade level is kept in global value *required*), then, the learning style of that learner should be changed to its alternative from Information reception domain: *Visual* learning style. That implies that in the next session, the learner will be presented with resources that are defined to support a particular learning style category.

Details of the adaptation process implemented in Protus 2.1 and executed adaptation rules are presented in Sect. 9.3 (adaptation rules: from AR1 to AR22).

11.2.2 Calculation of Initial Learning Styles

Each question in the questionnaire belongs to one of the four dimensions of learning styles (*information processing, information perception, information reception* and *information understanding*). Eleven questions were assigned to each of the four dimensions. Two answers are offered for every question and each of them pulls the final result in one of two categories within the proper dimension of learning styles.

Once a learner submits answers, the system fills in appropriate table (Table 11.1) by entering the one point for each answer in the appropriate field. Then numbers

Table 11.1 Example of completed table for the learning styles determination

Active/Reflexive			Sensitive/Intuitive			Visual/Verbal			Global/Sequential		
Question	a	b	Question	a	b	Question	a	b	Question	a	b
1	1	–	2	1	–	3	–	1	4	1	–
5	–	1	6	1	–	7	1	–	8	–	1
9	–	1	10	–	1	11	–	1	12	1	–
13	–	1	14	–	1	15	1	–	16	1	–
17	1	–	18	–	1	19	1	–	20	1	–
21	1	–	22	–	1	23	–	1	24	–	1
25	–	1	26	1	–	27	1	–	28	–	1
29	1	–	30	–	1	31	–	1	32	1	–
33	1	–	34	1	–	35	–	1	36	1	–
37	1	–	38	–	1	39	1	–	40	1	–
41	1	–	42	–	1	43	1	–	44	1	–
Overall (sum of marks within one column)											
Active/Reflexive			Sensitive/Intuitive			Visual/Verbal			Global/Sequential		
	a	b		a	b		a	b		a	b
Sum	7	4	Sum	4	7	Sum	6	5	Sum	8	3
Index of the particular style											
−2			2			−1			−3			

from the same column are summed. The final index is calculated based on the number of responses marked as A and B. Thus, the value of the index is −6 if all eleven responses was of type A, the index is −5 if there are ten types A responses and only one type B, the index is −4 if nine answers was type A and two answers were type B, and so on.

11.2.2.1 Explanation of Index

If the calculated index is in the range of −2–2 then learner is slightly leaning to one or another category within the dimension. If the index is −4, −3, 3, or 4, then the learner moderately leaning to one category and it will be easier for him/her to learn in an environment that favours this category. In the case that the index is −6, −5, 5, or 6, learner lean extremely to one category within scale and may have great difficulty if taught in an environment that supports the opposite category.

Next, the system calculates numerical value of learning styles for a learner and initializes the appropriate learner model. Initialization of the learner model plays an important role in defining the initial options for personalisation of the system.

Initialization of the learner model plays an important role in defining the initial personalization options in the system.

Listings 11.1, 11.2 and 11.3 presents segments of code that calculate the numerical value of learning styles based on the responses of learners and initialize the appropriate learning model.

```
protected void processRequest(HttpServletRequest request, HttpServletResponse response)
       throws ServletException, IOException {
    response.setContentType("text/html;charset=UTF-8");

    int odgovori[] = new int[44];
    Styles stil = new Styles();

    for (int i = 0; i < odgovori.length; i++) {
        try {
            if (request.getParameter("RadioGroup"+(i+1)).equals("radio1")) {
                odgovori[i]=0;
            } if (request.getParameter("RadioGroup"+(i+1)).equals("radio2")) {
                odgovori[i]=1;
            }
        } catch (java.lang.NullPointerException e) {
            stil.setPoruka("Mora se dati odgovor na sva pitanja");
            request.setAttribute("stil", stil);
            request.getRequestDispatcher("Questionnaire.jsp").forward(request, response);
        }

    }
```

Listing. 11.1 Reading of learner's answers

Listing. 11.2 Adding
answers by groups

```
int a = 0, b = 0, c = 0, d = 0;
for (int i = 0; i < odgovori.length; i++) {
    if (i%4==0) {
        if(odgovori[i]==1){
            a++;
        }
    }else if (i%4==1){
        if(odgovori[i]==1){
            b++;
        }
    }else if (i%4==2){
        if(odgovori[i]==1){
            c++;
        }
    }else if (i%4==3){
        if(odgovori[i]==1){
            d++;
        }
    }
}
```

Listing. 11.3 Final
calculation of learning styles
index

```
if (a>5) {
    stil.setProcessing(((2*a-11)+1)/2);
} else {
    stil.setProcessing(((2*a-11)-1)/2);
}
if (b>5) {
    stil.setPercepcion(((2*b-11)+1)/2);
} else {
    stil.setPercepcion(((2*b-11)-1)/2);
}
if (c>5) {
    stil.setReception(((2*c-11)+1)/2);
} else {
    stil.setReception(((2*c-11)-1)/2);
}
if (d>5) {
    stil.setUnderstanding(((2*d-11)+1)/2);
} else {
    stil.setUnderstanding(((2*d-11)-1)/2);
}
```

Protus 2.1 with a code segment in Listing 11.1 reads the learner's answers and fill in the appropriate array.

Responses that belong to a category of learning styles within each dimension are added up (Listing 11.2).

Based on these data, the final indexes for each of four dimensions of learning styles are calculated (Listing 11.3).

Based on the obtained data, Protus 2.1 collects information on initial learning style, updates the *Learner model ontology* and presents the results of a questionnaire as presented on the Fig. 11.3.

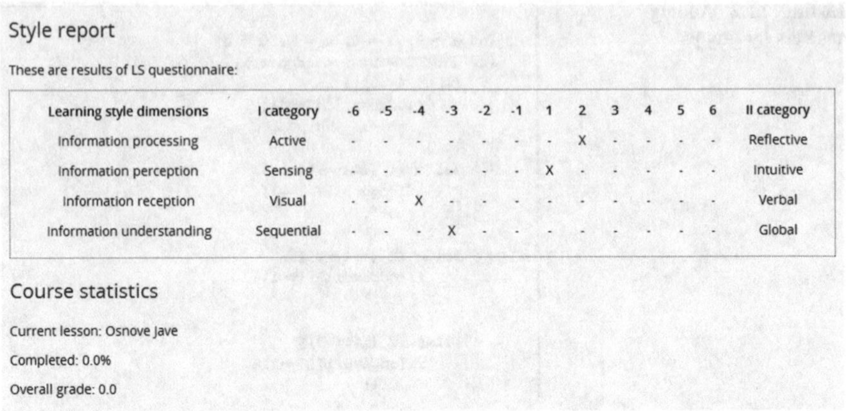

Style report

These are results of LS questionnaire:

Learning style dimensions	I category	-6	-5	-4	-3	-2	-1	1	2	3	4	5	6	II category
Information processing	Active	.	.	.	.	.	.	.	X	.	.	.	.	Reflective
Information perception	Sensing	.	.	.	.	.	.	X	.	.	.	.	.	Intuitive
Information reception	Visual	.	.	X	.	.	.	.	.	.	.	.	.	Verbal
Information understanding	Sequential	.	.	.	X	.	.	.	.	.	.	.	.	Global

Course statistics

Current lesson: Osnove Jave

Completed: 0.0%

Overall grade: 0.0

Fig. 11.3 Results of ILS questionnaire

11.2.3   Adaptation of User Interface Based on the Learning Styles

Protus 2.1 personalizes user interfaces to each individual learner, based on the learner model and their learning styles. Personalization includes presentation of recommended resources and links. Depending on the identified learning styles, Protus 2.1 modifies the user interface that presents teaching materials (Fig. 11.4). Process of user interface adaptation will be covered in next section, referring to all categories of learning styles separately.

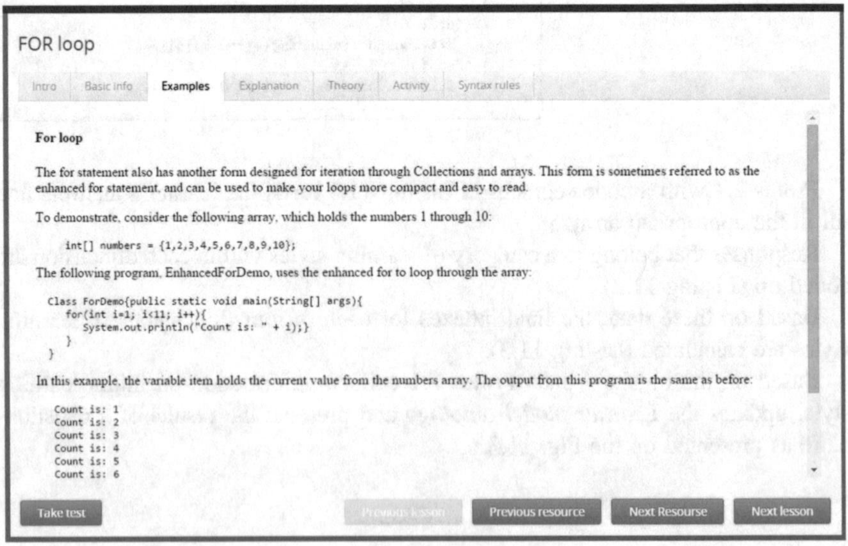

FOR loop

Intro   Basic info   **Examples**   Explanation   Theory   Activity   Syntax rules

For loop

The for statement also has another form designed for iteration through Collections and arrays. This form is sometimes referred to as the enhanced for statement, and can be used to make your loops more compact and easy to read.

To demonstrate, consider the following array, which holds the numbers 1 through 10:

```
int[] numbers = {1,2,3,4,5,6,7,8,9,10};
```

The following program, EnhancedForDemo, uses the enhanced for to loop through the array:

```
Class ForDemo{public static void main(String[] args){
    for(int i=1; i<11; i++){
        System.out.println("Count is: " + i);}
    }
}
```

In this example, the variable item holds the current value from the numbers array. The output from this program is the same as before:

```
Count is: 1
Count is: 2
Count is: 3
Count is: 4
Count is: 5
Count is: 6
```

Take test   Previous lesson   Previous resource   Next Resourse   Next lesson

Fig. 11.4 Java programming course in Protus 2.1

11.2.3.1   Information Processing: Active and Reflective Learners

Within Information *Processing domain*, we can distinguish example-oriented learners, called *Reflectors*, and activity-oriented learners, called *Activists* (Kolb 1984). Active learners tend to retain and understand information best by doing something active with it—discussing or applying it or explaining it to others. Reflectors are learners who tend to collect and analyse data before taking an action. They may be more interested in reviewing other learners' and professional opinions than doing real activities. In Protus 2.1 system, a learner with the active learning style is shown an activity first, then a theory, explanation and example (Fig. 11.5). For the learner with the reflective style this order is different—(s)he is shown an example first, then an explanation and theory, and finally (s)he is asked to perform an activity (Fig. 11.6).

11.2.3.2   Information Perception: Sensing and Intuitive Learners

Within Information *Perception domain*, sensing learners, called *Sensors*, tend to be patient with details and good at memorizing facts and doing hands-on (laboratory) work. On the other hand, intuitive learners, called *Intuitive* learners, may be better at grasping new concepts and are often more comfortable with abstractions and mathematical formulations than sensing learners. Sensors often prefer solving problems using well-established methods, and dislike complications and surprises. On the other hand, *Intuitive* learners like innovation and dislike repetition. Sensors tend to be more practical and careful than *Intuitive* learners. *Intuitive* learners tend to work faster and to be more innovative than Sensors. Presentation of the lesson to the learner with sensing and intuitive styles is given in Figs. 11.7 and 11.8. For

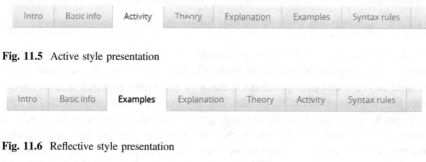

| Intro | Basic info | Activity | Theory | Explanation | Examples | Syntax rules |

Fig. 11.5   Active style presentation

| Intro | Basic info | Examples | Explanation | Theory | Activity | Syntax rules |

Fig. 11.6   Reflective style presentation

Fig. 11.7   Sensing style presentation

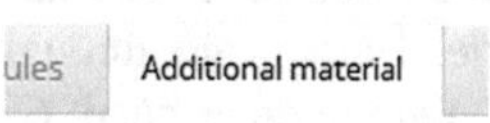

| ...ules | Additional material |

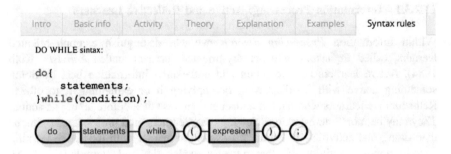

Fig. 11.8 Intuitive style presentation

example, it is assumed that sensing learners will be interested in additional materials, therefore they may click the button for additional material on the interface (Fig. 11.7). *Intuitive* learners are provided with abstract material, formulas and concepts as shown in Fig. 11.8. Adequate explanations are given in the form of block diagrams or exact syntax rules.

11.2.3.3 Information Reception: Visual and Verbal Learners

Within *Information Reception* domain, *Visual* learners remember best what they see —pictures, diagrams, flow charts, time lines, and demonstrations. *Verbal* learners get more out of words—written and spoken explanations. Figure 11.9 shows a presentation of the topic *For loop* to a learner with a preference for textual material (verbalizer style). Figure 11.10 shows the presentation of the same material to a learner with a visual preference. Based on the visual preference, the topic about the *For loop* is presented as a block diagram.

11.2.3.4 Information Understanding: Sequential and Global Learners

Within *Information Understanding* domain *Sequential* learners tend to follow logical stepwise paths in finding solutions. On the other hand, *Global* learners may be able to solve complex problems quickly or put things together in novel ways once they have grasped the big picture, but they may have difficulty explaining how they did it. Sequential learners prefer to go through the course step by step, in a linear way with each step following logically from the previous one, while global learners tend to learn in large leaps, sometimes skipping learning objects and jumping to more complex material. According to these characteristics of Sequential learning style, learners go through lessons in Protus 2.1 by a predefined order (Fig. 11.11). On the other hand, *Global* learners are provided with an overall view

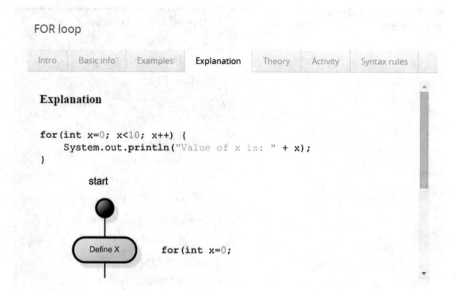

Fig. 11.9 Verbal style presentation

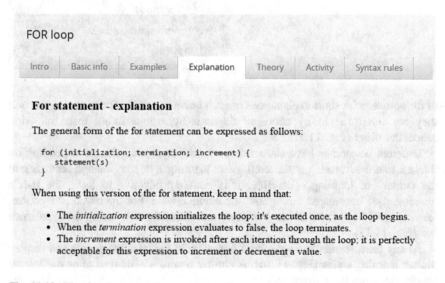

Fig. 11.10 Visual style presentation

Fig. 11.11 Navigation buttons for sequential learners

Fig. 11.12 Presentation for global learners

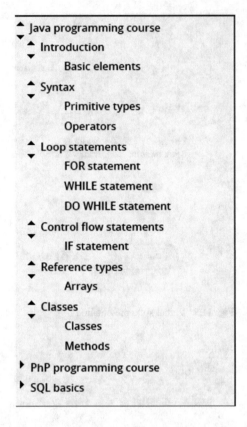

of the course, with short explanations of each unit and options for accessing the unit they are interested in by choosing the unit hyperlinks rather than following sequential order (Fig. 11.12).

Different researches have shown that learning style may change depending on the task that the learner has mastered. Also, learning style may change according to the content of learning. Therefore, it is counterproductive to leave the user's learning style unchanged throughout the whole course. For the rest of the course, learners were free to switch between presentations methods by using the experience bar (Fig. 11.13).

At any time, learner can get a brief overview of the attended course and display his/her learning styles (Fig. 11.14). A similar report is displayed after the learner checks out from the current session.

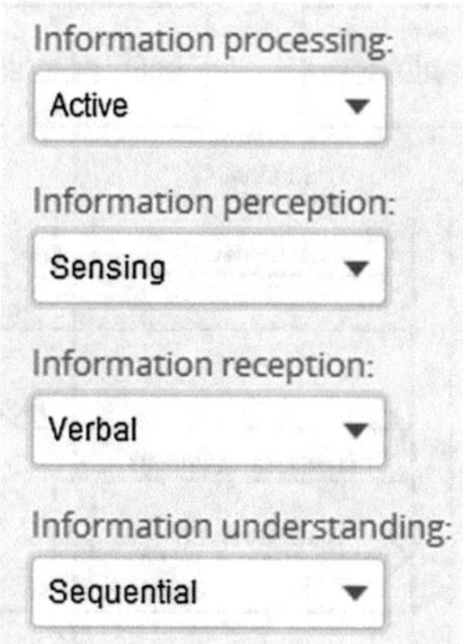

Fig. 11.13 Experience bar

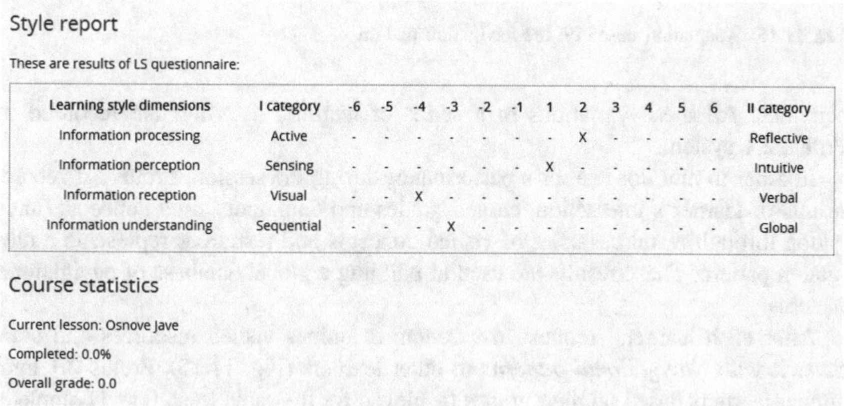
Style report

These are results of LS questionnaire:

Learning style dimensions	I category	-6	-5	-4	-3	-2	-1	1	2	3	4	5	6	II category
Information processing	Active	-	-	-	-	-	-	-	X	-	-	-	-	Reflective
Information perception	Sensing	-	-	-	-	-	-	X	-	-	-	-	-	Intuitive
Information reception	Visual	-	-	X	-	-	-	-	-	-	-	-	-	Verbal
Information understanding	Sequential	-	-	-	X	-	-	-	-	-	-	-	-	Global

Course statistics

Current lesson: Osnove Jave

Completed: 0.0%

Overall grade: 0.0

Fig. 11.14 Style report

11.3 Resource Sequencing

Resource sequencing is a well-established technique in application of intelligent tutoring systems in the educational process (Janssen et al. 2007). The idea of resource sequencing is to generate a personalized course for each learner by dynamically selecting the most optimal teaching actions, presentation, examples, task or problems at any given moment. By optimal teaching action, it is considered an operation that in the context of other available operations brings the learner closest to the ultimate learning goal. Most often, the goal is to learn and acquire some knowledge up to a specific level in an optimal amount of time. Learners could follow different paths based on their preferences and generate a variety of learning

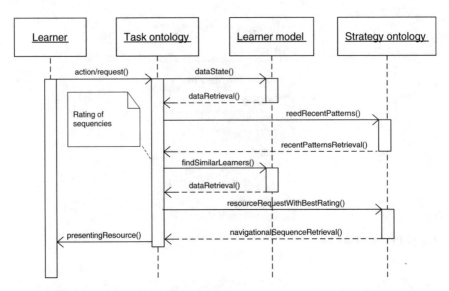

Fig. 11.15 Adaptation based on the navigation pattern

activities. All these variations in a series of learning activities are recorded by Protus 2.1 system.

In order to monitor learner's performance during the session, Protus 2.1 records results of learner's interaction, earned grades and data about used concepts (navigation through resources). Set of visited concepts and resources represents a navigation pattern. These results are used in building a global database of navigational patterns.

After each learners' request, the system examines visited resources and compares it with *Navigational patterns* of other learners (Fig. 11.15). Protus 2.1 finds similar learners based on their grades (achieved for the same lesson) and completes personalized recommendation of the learning content according to the ratings of these frequent sequences. Patterns are ranked based on the results achieved by learners on tests (Vesin et al. 2011). Specifically, the average score is calculated for learners that passed through the teaching material in the same order.

If the learner has successfully mastered a particular concept (Table 10.9, Sect. 9.3.1.), which is supported by specific resources (Table 10.10, Sect. 9.2), then the system adds that resource to the appropriate successful navigation pattern (Fig. 11.16).

Specifically, this would mean that if learner interacted with system through a certain navigation pattern, then system should recommend the next highest ranked resource that fits the pattern.

For example, let learner has visited resources that display introductory remarks (*Intro* resource), basic information (*basicInfo* resource) and various explanations (*Explanation* resource) for a lesson. While recommendations which resource will be presented to an active learner are being generated, the system takes into account learners who visited resources in the same order within the same lesson. The system

Fig. 11.16 User interface adaptation **a** Annotation **b** Presented interface elements **c** Adding elements

then compares the grades of those learners who afterwards visited *Syntax rules* resource and the grades of those learners who visited *Activity* resource. Resource, that brought higher scores to learners who have visited it, is being recommended to current learner.

Recommendation status of that resource is set to *true*, therefore, one of the several changes in user interface must be made (depending on the included resource type):

- link to that resource is annotated or highlighted (Fig. 11.16a),
- the interface elements for sequential navigation will be hidden, giving the learner possibility to freely jump through the courseware (in a case of sequential learning style category) or presented (in a case of sequential learning style category, Fig. 11.16b),
- additional tabbed pane elements will be added to related or more complex content to help situate the learnt subject and contribute in creating a clear overall view on the subject being thought (Fig. 11.16c).

Also, if learner visits a particular concept and during this interaction solve test related to this concept and obtain positive evaluation, then system records the results of this interaction in the learner model ontology (Listing 11.4) with adaptation rules presented in Sect. 9.3 (adaptation rules AR27-AR30) and marks that resource as visited (Table 10.10, Sect. 9.3).

11.3.1   Identification of Sequences of Learning Activities and Personalized Recommendation

In contrast to traditional classroom-based learning, the learning behaviour in Web-based environments is more determined by the learner's own decisions how to organize learning process (Northrup 2001). Learners could follow different paths based on their preferences and generate a variety of learning activities. All these variations in series of learning activities are noted down by Protus 2.1 system.

```
begin ProtusKurs
do
 presentConcept
 chooseFirstResource
 do
     if presentationEnd
         presentTest
         calculateResults
     else
         chooseNextResource
         presentResource
         adaptUserInterface
     end if
     updateSession
 while conceptIsNotMastered
while courseIsNotMastered
```

Listing. 11.4 Pseudo code of learning process in Protus 2.1

Listing. 11.5 Algorithm for pruning of sequential patterns

```
for all sequences   c∈ C_k  do
    for all (k-1)- subsequences s of c do
        if (s∉ L_{k-1})then
            delete c from C_k;
```

In order to investigate learning activities in detail, sequential pattern mining algorithm of AprioriAll (Pi-lian 2005) is adopted to extract behavioural (interaction) patterns from the log file. These patterns will be useful to analyse how learners evolve from the beginning of learning of particular unit, until they successfully finish it, or even give up. Learners with different learning styles have different sets of frequent sequences. Hence, learners were clustered based on their learning styles and then behavioural patterns were discovered for each learner by AprioriAll algorithm.

11.3.1.1   The Process of Mining Sequential Patterns by Apriori All Algorithm

Let $l = \{i_1, i_2, \ldots, i_m\}$ be a set of learning objects, called *items*. An *itemset* is a non-empty set of *items*. A sequence is an ordered list of *itemsets*. We denote an *itemset* i by $i_1, i_2, \ldots, i_m$, where i_j is an item. We denote a sequence s by $(s_1, s_2, \ldots, s_n)$, where s_j is an itemset. A sequence $\langle a_1, a_2, \ldots, a_n \rangle$ is contained in another sequence $\langle b_1, b_2, \ldots, b_n \rangle$ if there exist integers $\langle k_1 < k_2 < \ldots < k_n \rangle$ such that $a_1 \subseteq b_{k_1}, a_2 \subseteq b_{k_2}, \ldots, a_n \subseteq b_{k_n}$.

In a set of sequences, a sequence s—is *maximal* if s is not contained in any other sequence (Agrawal and Srikant 1995). A learner supports a sequence s if s is contained in the learner-sequence for this learner. The support for sequence is defined as the fraction of total learners who support this sequence.

Given a database D of learners' access transactions, the problem of mining sequential patterns is to find the maximal sequences among all sequences that have

Table 11.2 Database sorted by learner-id and transaction time

Learner-id	Access-time	Access path
1	2010. January 20.	Lesson1 introduction, Lesson1 overview, Lesson 1 theory
1	2010. January 22.	session Lesson1 exercise, Lesson 1 syntax rule,
1	2010. January 23.	Lesson 1 example 2, Lesson 1 example 3
1	2010. January 24.	Lesson 1 example 1, Test 1
2	2010. January 15.	Lesson 1 overview, Lesson 1 theory session
2	2010. January 16.	Lesson 1 example 3, Lesson 1 example 1,
2	2010. January 17.	Test 1
3	2010. January 18.	Lesson 1 theory session
3	2010. January 20.	Lesson 1 syntax rule
3	2010. January 21.	Lesson 1 example 3, Lesson 1 example 1
3	2010. January 22.	Lesson 1 exercise, Test 1
4	2010. January 21.	Lesson1 introduction, Lesson1 overview, Lesson 1 theory
4	2010. January 23.	session, Lesson1 exercise,
4	2010. January 24.	Lesson 1 syntax rule,
4	2010. January 26.	Lesson 1 example 2, Lesson 1 example 3, Test 1
5	2010. January 16.	Lesson1 introduction, Lesson1 overview
5	2010. January 17.	Lesson 1 theory session
5	2010. January 18.	Lesson1 exercise
5	2010. January 19.	Lesson 1 example 1, Lesson 1 example 2, Test 1

a certain learner-specified minimal support. Each such maximal sequence represents a *sequential pattern*. We call a sequence satisfying the minimum support constraint a *large* sequence.

The process of mining sequential patterns can be split into five phases (Agrawal and Srikant 1995). To conveniently explain them, we use a small part of the database, as shown in Table 11.2. Each transaction consists of the following fields: learner-id, access-time, and the access path in the transactions. Phases are:

- *Sort phase*: The original database is sorted with learner-id as the major key and access time as the minor key. Table 11.2 shows the result set of learner sequences after sorting.
- *Large-itemsets (l-itemsets) phase*: In this phase, we find the set of all large itemsets. Without loss of generality, we assume that the set of l-items is mapped to a set of consecutive letters. Suppose the minimal support is 60 %, and the minimal support customer sequence is thus 3. The result of large 1-itemsets is listed in Table 11.3.
- *Transformation phase*: In a transformed learner sequence, each transaction is replaced by the set of l-itemsets contained in that transaction. If a transaction does not contain any l-itemset, it is not retained in the transformed database. This transformed database is shown in Table 11.4.
- *Sequence phase*: This is an essential phase of the process. In this phase, an algorithm uses a set of large itemsets to find the desired sequences. The idea is that, given the l-itemsets, the set of all the sequences with minimum support should be found. In each pass, we use the large sequences of the previous pass to generate the candidate sequences, and then measure their support by making a

Table 11.3 Large itemsets

Large itemsets	Mapped to
Lesson1 introduction	a
Lesson1 overview	b
Lesson 1 theory session	c
Lesson1 exercise	d
Lesson 1 syntax rule	e
Lesson 1 example 1	f
Lesson 1 example 2	g
Lesson 1 example 3	h
Test 1	i

Table 11.4 Transformed database

Learner id	Mapped to
1	<(abc)(de)(gh)(fi)>
2	<(bc)(hf)i>
3	<ce(hf)(di)
4	<(abc)de(ghi)
5	<(ab)cd(fgi)>

pass over the database. The first pass over the database is made in the 1-itemset phase, and we determine the large 1-sequences shown in Table 11.3. The large sequences together with their support at the end of the third and fourth pass are shown in Tables 11.5 and 11.6, respectively.

- *Maximal phase*: to reduce information redundancy, the sequential patterns contained in other sequential patterns are pruned (see algorithm in Listing 11.1). Table 11.7 shows Maximal Large 5-Sequences, after pruning.

11.4 Recommendation Process Based on Collaborative Filtering

The task of a collaborative filtering system is to predict the usefulness rating of a particular learner l for a similar learner l' (Herlocker et al. 2004). Therefore, the rating vector of a learner l is represented by $R_l = (r_{l1}, r_{l2}, ..., r_{li})$. The entry r_{lj} of R_l is provided by Protus 2.1 to indicate the learner's knowledge degree for the unit (s) he is currently used in the learning process.

The collaborative filtering system compares the learner's ratings with the ratings of all other learners, who have been rated. Then a weighted average of the other learners rating is used as a prediction. If S_l is set of frequent sequences that a learner l has been rated for, then we can define the mean rating of learner l as:

Table 11.5 Large 3-sequences

Sequence	Support
abc	3
abd	3
abg	3
abi	3
abh	3
bcd	3
bce	3
bcg	3
bch	3
bcf	3
bdg	3
bdh	3
bgh	3
bhi	4
bci	4
ceh	4
cdg	3
cdi	4
chi	3
ghi	3
hfi	3
ehf	3
dgi	3
dgh	3

Table 11.6 Large 4-sequences

Sequence	Support
abcd	3
abcg	3
abch	3
abgh	3
abhi	3
abdg	3
abdh	3
bcgh	3
bchi	3
bghi	3
cdgh	3
cehf	3
cdgi	3
dghi	3

Table 11.7 Maximal large
5-Sequences (after pruning)

Sequences
\<abcgh\>
\<abchi\>
\<abghi\>

$$\bar{r}_l = (1/|S_l|) \sum_{i \in S_l} r_{li}$$

When Pearson correlation (Herlocker et al. 1999) is used, similarity is determined from the correlation of the rating vectors of learner l and the other learner l'. This value measures the similarity between the two learners' rating vectors.

$$\rho(l,l') = \left(\sum_{i \in S_l \cap S_{l'}} (r_{li} - \bar{r}_l) \cdot (r_{l'i} - \bar{r}_{l'}) \right) \Big/ \left(\sqrt{\sum_{i \in S_l \cap S_{l'}} (r_{li} - \bar{r}_l)^2 \cdot \sum_{i \in S_l \cap S_{l'}} (r_{l'i} - \bar{r}_{l'})^2} \right)$$

The prediction formula is based on the assumption that the prediction is a weighted average of the other learners' ratings.

$$p^{col}(l,i) = \bar{r}_l + k_{li} \sum_{l \in L_i} \rho(l,l')(r_{l'i} - \bar{r}_{l'})$$

where L_i—is the set of learners who were rated for sequence i; the factor k_{li} is used to normalize the weights.

$$k_{li} = 1 \Big/ \left(\sum_{l' \in L_i} \rho(l,l') \right)$$

When this procedure is executed, Protus 2.1 can recommend relevant links and actions to target learner during the learning process based on similarities with other learners. The system can be considered successful if the observed learner is rated with a similar grade.

11.5   Tag-Based Personalized Recommendation Using Ranking with Tensor Factorization Technique

The task of tag-based personalized recommendation is to provide a learner with a personalized ranked list of tags for a specific item. We have implemented the recommendation component of Protus 2.1 system that recommends the most popular tags and experimentally compare it with the previous version of the system, which will be shown in Chap. 5. On the basis of comprehensive comparisons of techniques that can be used to recommend tags, in the rest of this section, we will show the possibility of implementing the system using Ranking with Tensor Factorization (*RTF*) technique, as analyzed in the Sect. 7.4.4. The recommendation process consists of three phases:

- generating initial tensor,
- computing tensor factorization,
- generating a list of recommended items.

11.5.1   Generating Initial Tensor

To generate the initial tensor, we have been used 3-dimensional data of learners, items (learning objects) and tags. The third-order tensor $\mathcal{A} \in R^{I \times J \times K}$ represents this data where I, J and K are the dimensions of the data of learners, items and tags, respectively. A value $(\mathcal{A})_{ijk} = a_{ijk}$ can represent, for example, how many times learner i tagged an item k with a tag j. In this phase following steps can be recognized:

1. A learner set is generated.
2. Set of tags is resolved. These are the tags used by the learners.
3. Item set is resolved. These are the items tagged with the tags by the learners.
4. Iterate through all the learners, tags and items. Resolve if a current item is tagged by the current tag and learner. If so—mark the existing relationship in the tensor.
5. Store the empty relations (if a learner does not tag an item with a tag) of a current learner. These relations will be used to resolve the recommendations.

11.5.2   Computing Tensor Factorization

To compute a tensor factorization, the initial tensor has to be defined, and then the following steps should be applied:

1. Firstly, the initial tensor is split into the three mode matrices.
2. Secondly, the dimensions are reduced for each mode matrix. These reduced matrices are multiplied to compute a core tensor.
3. Finally, the reduced matrices are transformed, multiplications are applied. The factorized tensor is computed.

11.5.3 Generating a List of Recommended Items

When the factorized tensor is computed, the recommendations can be determined. The task of tag recommendation is to predict which tags a learner is most likely to use for tagging an item. That means a tag recommender has to predict the numerical values of the factorized tensor indicating how much the learner likes a tag for an item. Instead of predicting single elements the system should provide the learner a personalized list of the best N tags for the item.

11.5.4 Tag-Based Recommendation in Protus 2.1

Collaborative tagging activities in e-commerce caused the appearance of tag-based user's profiling approaches, which assume that users expose their preferences for certain contents through tag assignments (Manouselis et al. 2011). Thus, the tags could be interesting and useful information to enhance recommender system's algorithms. The innovation with respect to the e-learning system lies in their ability to support learners in their own learning path by recommending tags and learning items, and also their ability to promote the learning performance of individual learners.

A tag is a keyword assigned by a user to represent the subject content, format, utility or affective characteristics of a bookmark, photograph, video, audio, post, wiki, blog or other online resources. The goal of tagging is to make a collection of resources easier to search, to discover, to share and to navigate (Ding et al. 2008). Using tags for characterizing digital educational resources is commonly referred to as collaborative tagging, whereas the collection of tags created by the different users individually is referred to as folksonomy (Zervas and Sampson 2013).

Learners could benefit from writing tags in several important ways. Tagging is proven to be a meta-cognitive strategy that involves learners in active learning and engages them more effectively in the learning process. As summarized by (Bonifazi et al. 2002), tags could help learners to remember better by highlighting the most significant part of a text, could encourage learners to think when they add more ideas to what they are reading, and could help learners to clarify and make sense of the learning content while they try to reshape the information. Learners' tags could create an important trail for other learners to follow by recording their thoughts about specific learning material and could give more comprehensible recommendations about the learning process. Tags presented on a webpage can give a learner some idea of its importance and its content.

The information provided by tags makes available insight on learner's comprehension and activity, which is useful for both learners and teachers. Tagging, by its very nature, is a reflective practice which can give learners an opportunity to summarize new ideas, while receiving peer support through viewing other learners' tags/tag suggestions. Tagging interface in Protus 2.1 (Fig. 11.17) provides possible

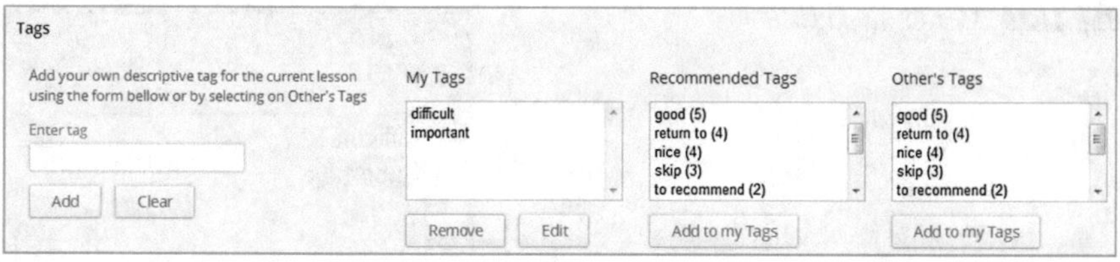

Tags			
Add your own descriptive tag for the current lesson using the form bellow or by selecting on Other's Tags	My Tags	Recommended Tags	Other's Tags
Enter tag	difficult important	good (5) return to (4) nice (4) skip (3) to recommend (2)	good (5) return to (4) nice (4) skip (3) to recommend (2)
Add Clear	Remove Edit	Add to my Tags	Add to my Tags

Fig. 11.17 Tags menu

solutions for learners' engagement in a number of different annotation activities—add comments, corrections, links, or shared discussion.

Learning resources in Protus 2.1 have been created using the authoring tools by instructors and they have been stored in a resource repository. Along presentation of resources to learner, user interface in Protus 2.1 also contain options for creating and reviewing tags for every resource. Therefore, learners are able to view and rate the resources.

To create a tag in Protus 2.1 the learner simply starts by choosing an active learning object in the content and enter arbitrary keywords in the appropriate *textfield* (Fig. 11.18). The system allows participants to enter as many tags as they wish, separated by commas. This makes it possible to use spaces in tags, rather than restricting the participant to a single word. This is in contrast to many popular tagging systems which only allow single word tags. The system allows multi-word tags to eliminate the problem of establishing a convention for word combination.

Whenever the learner returns to that particular learning object, the list of tags (s) he has previously made will re-appear (Fig. 11.19). When particular tag is selected, two options are presented: *Edit* and *Remove*, which give learner option to modify or delete this tag.

The most popular tags added by other learners, appear under *Others' Tags*. Learner has ability to add any tags from the *Others' tags* to *My tags* list. An example of these functionalities is shown in Fig. 11.20.

Fig. 11.18 Interface for creating tags

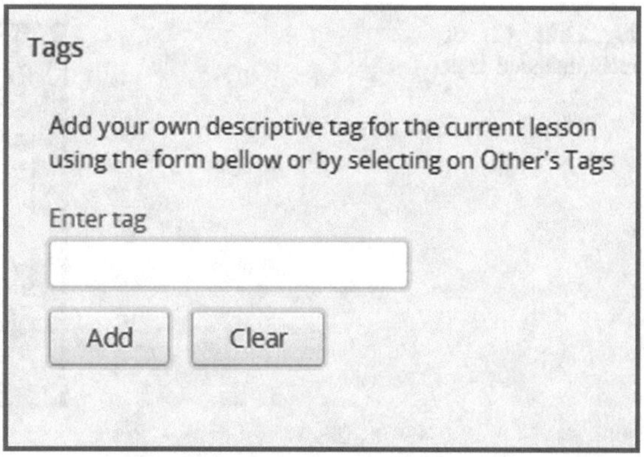

Tags

Add your own descriptive tag for the current lesson using the form bellow or by selecting on Other's Tags

Enter tag

Add   Clear

Fig. 11.19 List of *my tags*

Fig. 11.20 List of other's tags

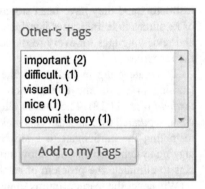

According to the research that we have made by comparative analysis of tag-based recommender algorithms, the *recommended tags* list is generated according to the learners' and experts' tags based on *Ranking with Tensor Factorization model* which produced more accurate recommendations then existing state-of-the-art algorithms (Milicevic et al. 2010) (Fig. 11.21).

Fig. 11.21 List of recommended tags

11.5.4.1 Tag Browsing

Tag browsing, in terms of an individual learner, is an interface that automatically categorize information based on tags. In the community sense it is a way to gain a *global view* of the tagging of the entire community, while still allowing learners to browse the individual contributions of peers. This functionality can be accessed through the tag menu, and provides three options: *My Tags*, *Others' tags*, and *Community Tags*.

Information provided by the individual learner is located under *My Tags* in the interface (Fig. 11.22). *My Tags* list presents all the tags the learner has been used, which are ordered from the most to least frequently used tag. When individual tag is selected, an option *Visit selected lesson* is presented, which links the learner to the lesson that was tagged. Textfield named *Name filter* adds functionalities for search among tags.

By expanding the *Others' Tags* section, the list of active learners in community and the tags they entered (Fig. 11.23).

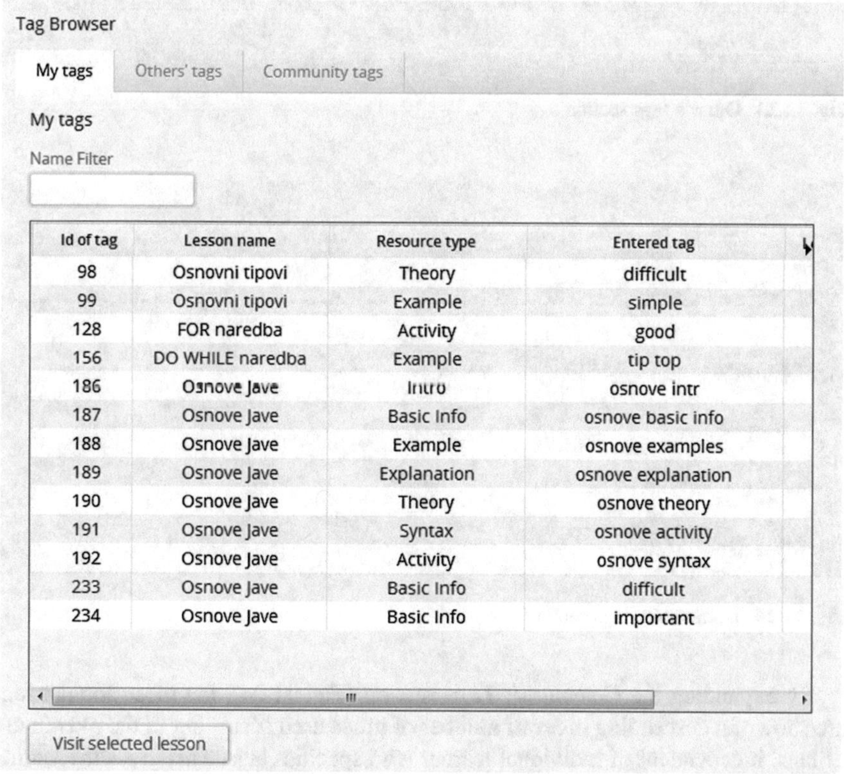

Tag Browser

| My tags | Others' tags | Community tags |

My tags

Name Filter

Id of tag	Lesson name	Resource type	Entered tag
98	Osnovni tipovi	Theory	difficult
99	Osnovni tipovi	Example	simple
128	FOR naredba	Activity	good
156	DO WHILE naredba	Example	tip top
186	Osnove Jave	Intro	osnove intr
187	Osnove Jave	Basic Info	osnove basic info
188	Osnove Jave	Example	osnove examples
189	Osnove Jave	Explanation	osnove explanation
190	Osnove Jave	Theory	osnove theory
191	Osnove Jave	Syntax	osnove activity
192	Osnove Jave	Activity	osnove syntax
233	Osnove Jave	Basic Info	difficult
234	Osnove Jave	Basic Info	important

Visit selected lesson

Fig. 11.22 *My tags* section

Tag Browser

| My tags | **Others' tags** | Community tags |

Others' tags

Learner	Id of tag	Lesson name	Resource type	Entered tag
Bogunović Ana	192	Osnove Jave	Activity	osnove syntax
Bogunović Ana	98	Osnovni tipovi	Theory	difficult
Bogunović Ana	188	Osnove Jave	Example	osnove examples
Car Slobodan	97	Osnovni tipovi	Explanation	visual
Car Slobodan	100	Operatori	Intro	nice
Car Slobodan	53	Osnove Jave	Basic Info	good
Car Slobodan	213	Operatori	Theory	go back
Car Slobodan	155	DO WHILE naredba	Syntax	repeated
Car Slobodan	91	Operatori	Theory	return to
Car Slobodan	62	Osnove Jave	Basic Info	good
Car Slobodan	120	IF naredba	Theory	bing bag theory
driov Maja	220	Osnove Jave	Basic Info	good
driov Maja	223	Osnove Jave	Theory	osnove theory
driov Maja	219	Osnove Jave	Intro	osnove intr
driov Maja	222	Osnove Jave	Explanation	osnove explanation

Visit selected lesson

Fig. 11.23 Other's tags section

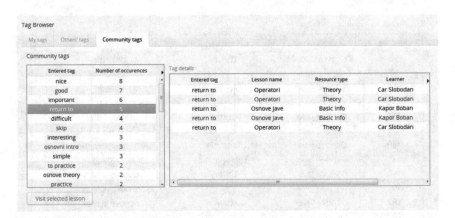

Fig. 11.24 Community tags section

By expanding the *Community Tags* section (Fig. 11.24), the most popular tags are shown in descending order of number of times used according to the overall use of tags, independent of individual learner who specifies tags. This gives the learner an idea, at a much higher level, the overall view of all the content. The learner can get a sense of what are the most important terms and/or ideas at a course level. By choosing one of the tags, the learner can visit lesson that was tagged. If a learner chooses to search for any specific tag, corresponding lesson and learning object will be displayed.

11.6 Use and Functioning of the System

Protus 2.1 is aims at developing courses in different domains. Java programming course is at this moment only fully developed course and used in real educational setting.

Protus 2.1 offers a Web interface for presenting of learning material and testing knowledge.

This section describes the user interface and explains the guidelines that were taken into account for its design. First screen after starting the system offers the possibility of registering a new user—option *Sign up* (Fig. 11.25) or signing in an existing one—option *Sign in* (Fig. 11.26). Each profile stores personal information supplied directly by the learner, like: last name, first name, birth year, login details, etc. (static information), and information about progress through course, grades, and learning styles (dynamic information). Dynamic data is constantly updated by the system during learning sessions.

A standardized form of the learner model presented in this monograph enables the exchange of learner's data with other systems.

The first time that learners use Protus *2.1,* system asks them to fill the questionnaire that contain the ILS questions to predict their own learning styles (Fig. 11.27).

Initial learning style of learner is calculated based on the learners' answers based on the learning style-model by Felder and Soloman. Learning style is recorded in appropriate learner model (Klašnja-Milićević et al. 2011b).

It is very important for learners to take questionnaire seriously and answer all provided questions because initialization of learner model is directly dependent on

Fig. 11.25 Sign up form of Protus 2.1

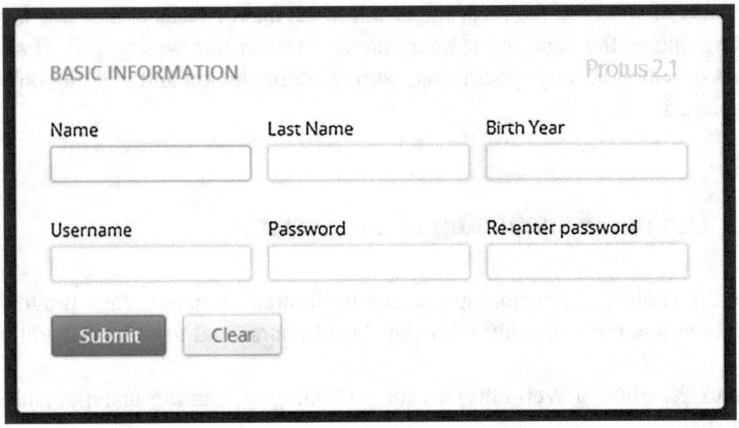

Fig. 11.26 Registration form of Protus 2.1

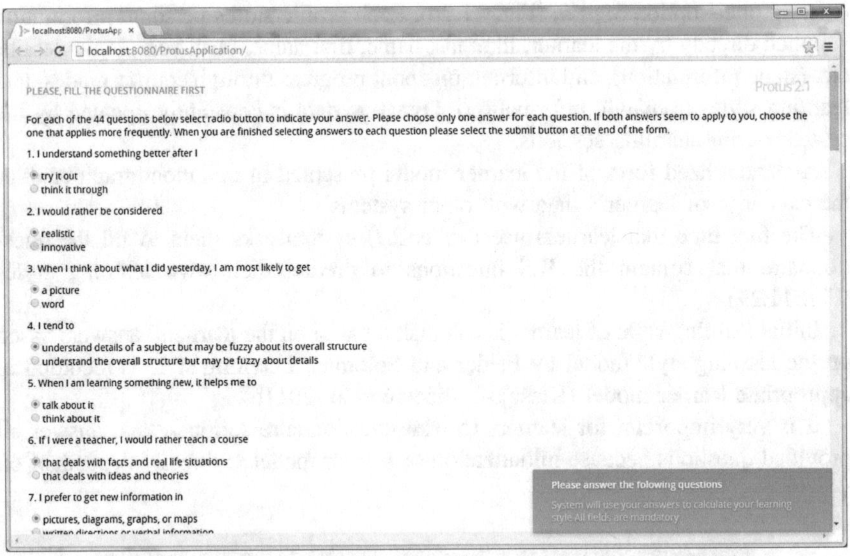

Fig. 11.27 ILS questionnaire

their answers. Protus 2.1 disables the start of the course until learner answers all the questions.

When a learner is logged in, a session is initiated based on learner's specific data and sequence of lessons is recommended to him/her. All information on the current lesson, percentage of mastered materials, learner's grades achieved during session, overall average grade, etc. are set to their initial values and entered in *Learner model ontology*. In this way, the learner begins chosen course (Fig. 11.28).

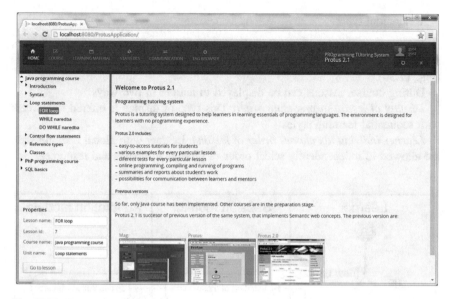

Fig. 11.28   Home page of Protus 2.1

11.6.1   Integration of Java Programming Course in Protus 2.1

Teaching material for Java programming course in Protus 2.1 is organized into six units with a total of 18 lessons (Fig. 11.29). Lessons are arranged in the following units:

- Introduction,
- Syntax,
- Loop statements,
- Execution control,
- Classes,

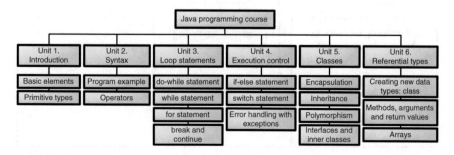

Fig. 11.29   Lessons hierarchy in Java programming course

- Referential types.

When a learner is logged in, a session is initiated based on learner's specific data from *Learner model ontology* and sequence of lessons is recommended to him/her (Fig. 11.30).

During course, lessons can be displayed to learner in two ways:

Display of lesson in predefine order. This type of display is offered to learners with sequential learning styles.

Learner individually choose order of lessons. Learner with global learning style are allowed to independently select order of presented lessons and resources.

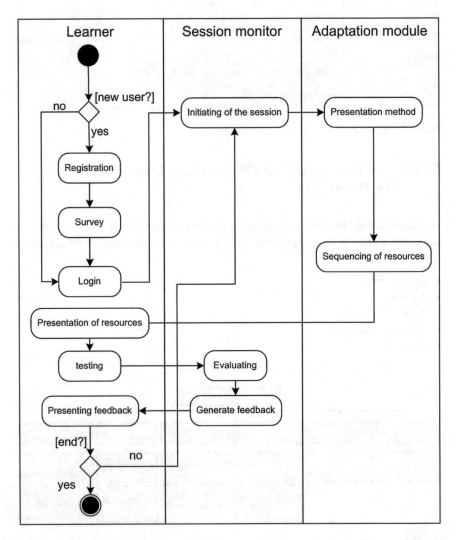

Fig. 11.30 Learning process in Protus 2.1

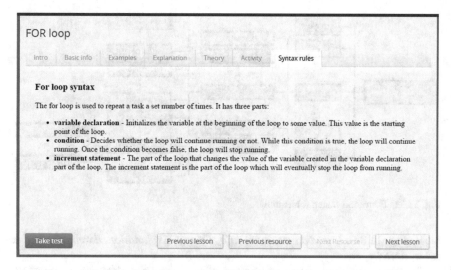

Fig. 11.31 Learner's interfaces

After selecting a lesson, from the collection of lessons available in Protus 2.1 (Fig. 11.31), system chooses presentation method of lesson based on the learner's preferred style. For the rest of the lesson, learners were free to switch between presentation methods by using the media experience bar.

User interface for learners is presented in Fig. 11.31. Each lesson is presented in a tabbed pane. Each tab contains corresponding part of the lesson, i.e. presents corresponding teaching resource.

Depending on the generated recommendations, order of tabs is changed, individual options are displayed/hidden, appropriate types of resource are displayed or certain links are highlighted.

11.7   Educational Material in Protus 2.1

This particular prototype of Protus 2.1 is realized to help learners in learning the essentials of Java programming languages (Klašnja-Milićević et al. 2011a). Java was chosen because it is a programming language widely used at our University, and because it is a clear example of an object-oriented language and therefore suitable for teaching the concepts of object-orientation.

The environment is designed for learners with no programming experience. It is an interactive system that allows learners to use the teaching material prepared within appropriate course. It also includes a part for testing the learner's acquired knowledge.

The learning content is divided into units, each of which consists of several lessons (*Concepts*). Every lesson contains several resources (presented in different

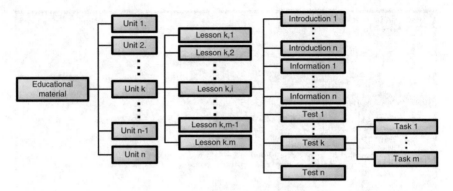

Fig. 11.32 Education material hierarchy

tabs—Fig. 11.32): *Introduction, Basic info, Theory, Explanation, Examples, Syntax rules, Activity*, etc.

To every lesson an unlimited number of resources and tests can be attached. Their number can be increased by teachers using an appropriate authoring tool. Resources that present explanations and syntax rules have two form of presentation —verbal and textual. Appropriate form is elected based on current category of learning styles within an *Information reception* domain for current learner.

Teaching materials in Protus 2.1 are presented in HTML form. System loads a recommended document for current learner and displays it in a Web browser.

Lessons (concepts) implemented in Java programming course are grouped into 6 units as shown in Fig. 11.33.

In the next Figures we will illustrate different recourses for some lessons. For example, the resources for *For loop* lesson illustrate structure and design of resource and its form of presentation to learner. Following resources are presented: *Introduction* (Fig. 11.34), *Basic information* (Fig. 11.35), *Examples* (Fig. 11.36), *Explanation* (Fig. 11.37), *Syntax rules* for learners with *verbal* learning style (Fig. 11.38) and *Syntax rules* for learners with *visual* style learning (Fig. 11.39).

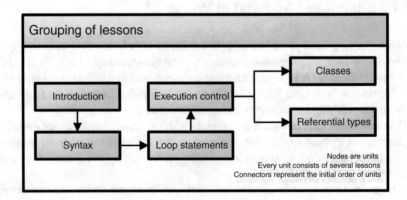

Fig. 11.33 Grouping of lessons

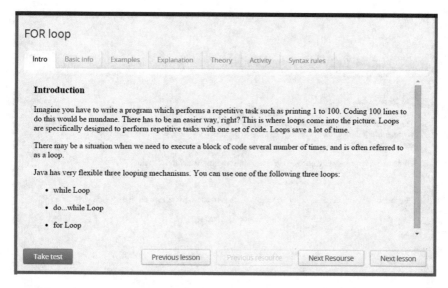

Fig. 11.34 Introduction in *For loop* lesson

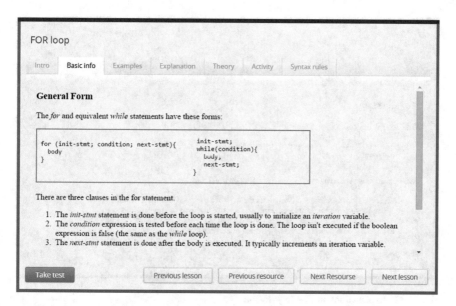

Fig. 11.35 Basic information in *For loop* lesson

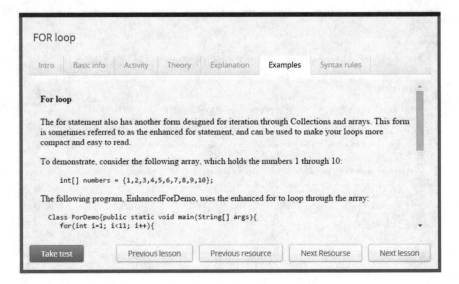

Fig. 11.36 Examples in *For loop* lesson

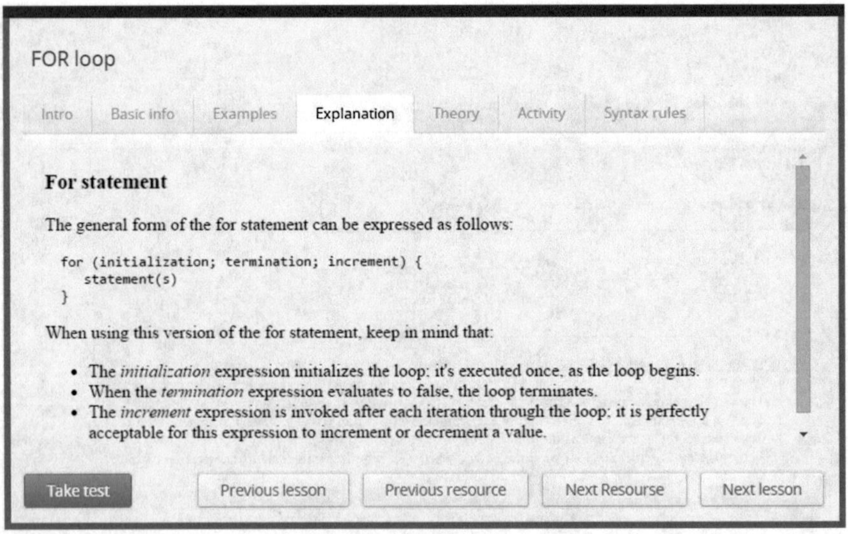

Fig. 11.37 Explanations in *For loop* lesson

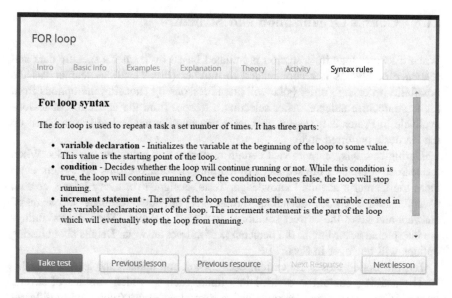

Fig. 11.38 Syntax rules for *Verbal* learner in *For loop* lesson

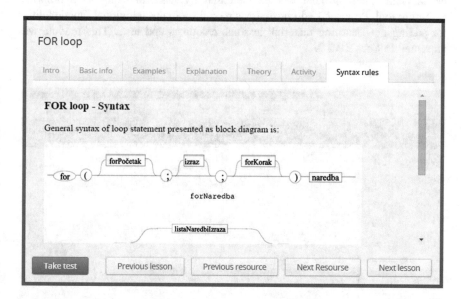

Fig. 11.39 Syntax rules for *Visual* learner in *For loop* lesson

11.8   Course Organization and Structure

When a learner logs in, a session is initiated based on learner's specific data and sequence of lessons is recommended to him/her (Fig. 11.40). A learner has the possibility to change order (s)he will attend lessons by choosing the options from the lesson/course sidebar. After selecting a lesson, from the collection of lessons available in Protus 2.1, system chooses a presentation method of lesson based on the learner's preferred style.

During sessions, learners visit certain resources and solve various tasks. When the learner completes the sequence of learning materials, Protus 2.1 system evaluates the learner's acquired knowledge. Tests, designed for every lesson, contain several multiple-choice questions. Protus 2.1 provides feedback on learners' answers and gives the correct solutions after every question. The learners' ratings are interpreted according to the percentage of correct answers. Details about testing options will be cover in Sect. 11.8.

Log out option closes the current user session and updates the learner model.

The learning content is divided into units, each of which consists of several lessons (*Concepts*). Every lesson contains several resources (presented in different tabs—Fig. 11.41): *Introduction*, *Basic info*, *Theory*, *Explanation*, *Examples*, *Syntax rules*, *Activity*, etc. To every lesson an unlimited number of resources and tests can be attached. Their number can be increased by teachers using an appropriate authoring tool. Protus 2.1 administrator's module contains additional functionalities for adding new learning material: lessons, resources and tests. This module was presented in Sect. 10.3.2.

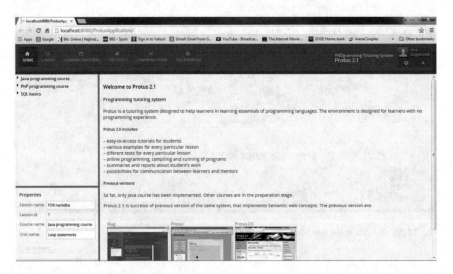

Fig. 11.40 Protus 2.1 welcome screen

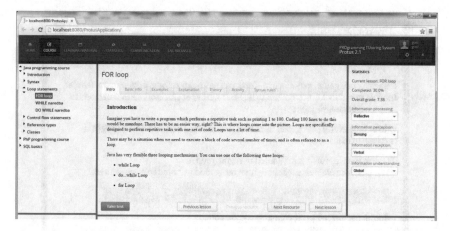

Fig. 11.41 Course options within Protus 2.1

Lessons browser

Id of course	Name of the course		Id of unit	Name of the unit		Id of lessor	Name of the lesson
1	Java programming course		1	Introduction		7	FOR naredba
2	PhP programming course		2	Syntax		8	WHILE naredba
3	SQL basics		3	Loop statements		10	DO WHILE naredba
			4	Control flow statements			
			5	Reference types			
			6	Classes			

Courses   Units   Lessons

Visit lesson

Fig. 11.42 Overview of offered courses

The learner is also provided with an overview of offered courses (Fig. 11.42) and menu with additional options. This menu includes shortcuts to a short user manual (Fig. 11.43) and settings of the basic system options (Fig. 11.44). Learner is also provided with possibilities for sending (Fig. 11.45) and receiving (Fig. 11.46) messages to/from teacher or other learners.

11.8.1   Testing in Protus 2.1

At the end of each lesson a post-testing should be conducted (Fig. 11.47). When the learner completes the sequence of learning materials, the system evaluates the degree of acquired knowledge. The test contains several multiple-choice questions.

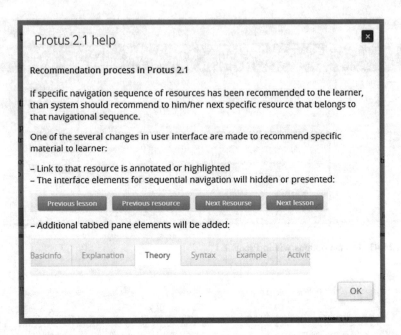

Fig. 11.43 Short manual

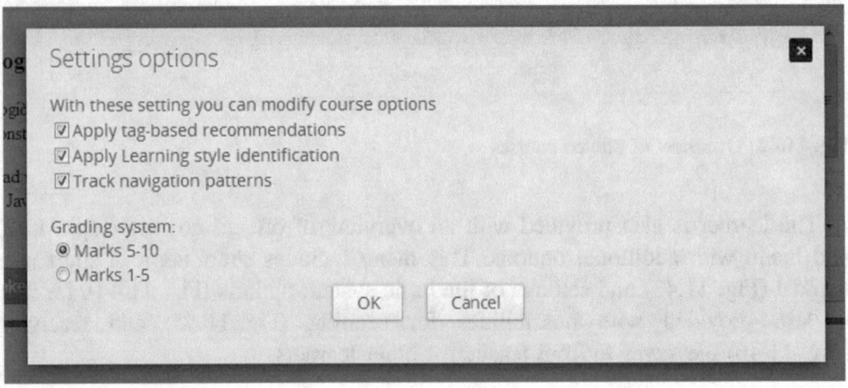

Fig. 11.44 Settings options

Protus 2.1 provides feedback on their answers and gives the correct solutions after every question (Figs. 11.48 and 11.49). The post-test section is followed by a test summary (Fig. 11.50).

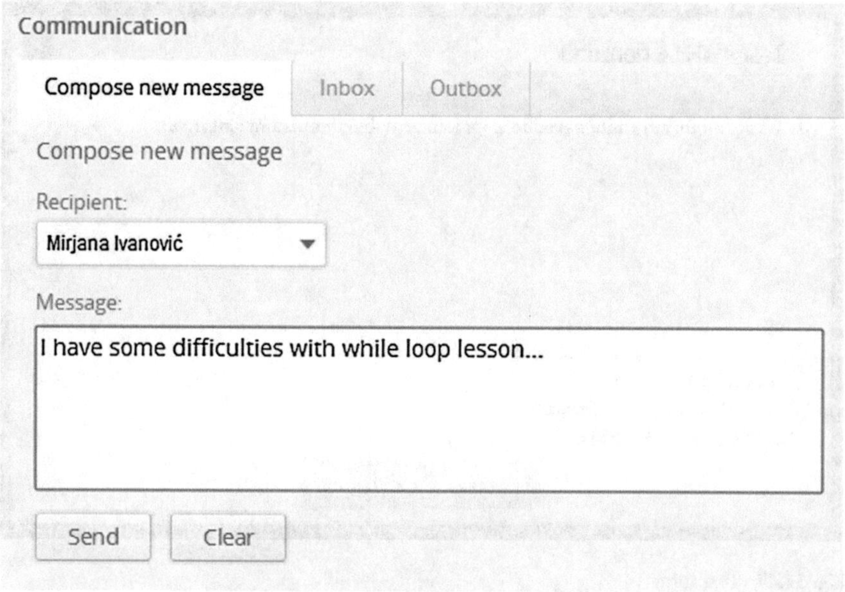

Fig. 11.45 Communication with other users

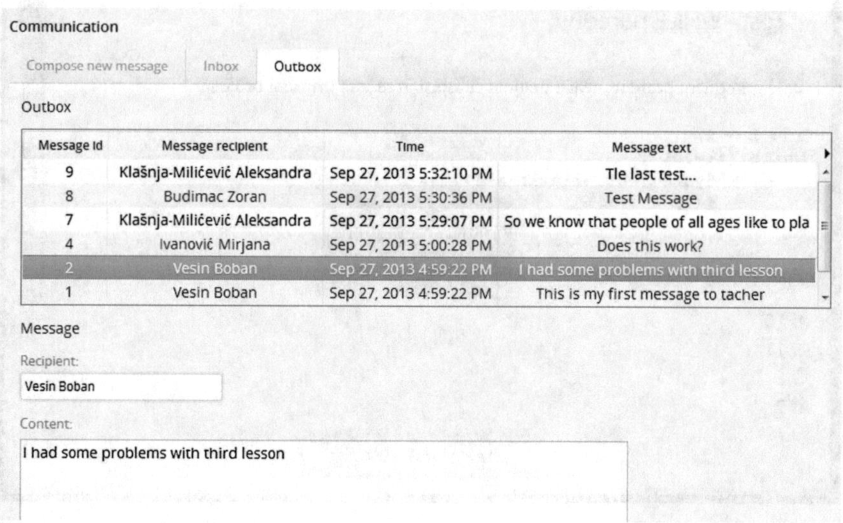

Fig. 11.46 Messages from other users

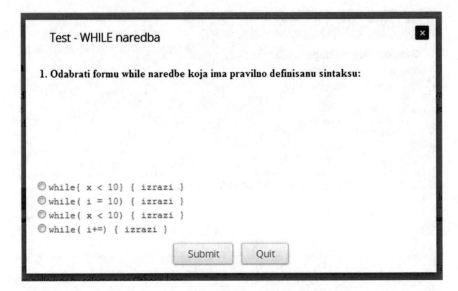

Fig. 11.47 Test form

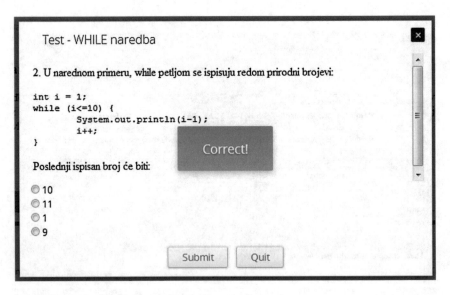

Fig. 11.48 Feedback after correct answer

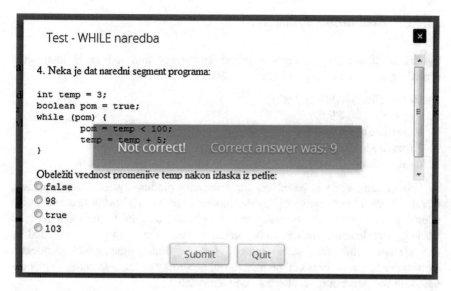

Fig. 11.49 Feedback after incorrect answer

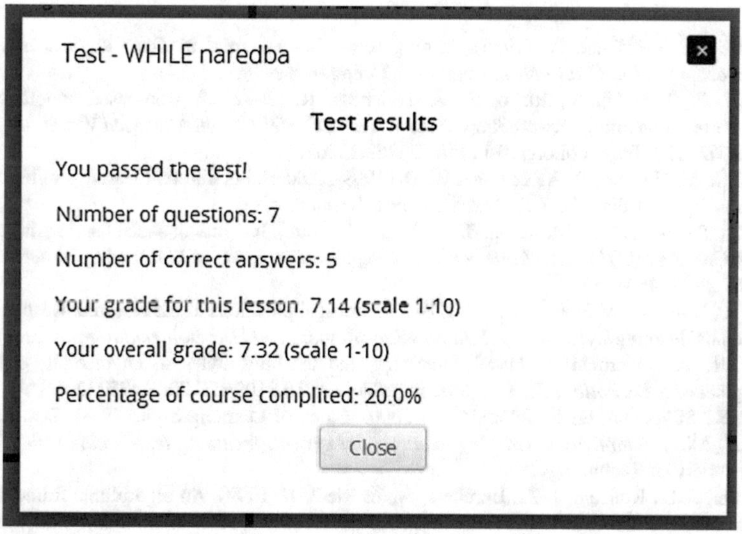

Fig. 11.50 Test summary

11.8.2   Evaluation Process

System evaluates the learner's acquired knowledge and ratings is interpreted according to the percentage of correct answers, as follows:

- 10 (excellent)—(90–100 %)
- 9 (good)—(80–89 %)
- 8 (average)—(70–79 %)
- 7 (passing)—(60–69 %)
- 6 (marginal)—(50–59 %)

This grading scale is based on our university grading system. Consequently, learners have a better sense of having mastered the material using this evaluation approach. The system can be easily transformed and adapted to other standards of grading. Two learners are said to be similar to each other if they are evaluated by the system with the same ratings for a similar navigation sequence. Recommendation process can be carried out according to these learning sequences based on the collaborative filtering (CF) approach.

References

Agrawal, R., & Srikant, R. (1995). Mining sequential patterns. In *Data Engineering, 1995. Proceedings of the Eleventh International Conference on* (pp. 3–14).

Bonifazi, F., Levialdi, S., Rizzo, P., & Trinchese, R. (2002). A web-based annotation tool supporting e-learning. *Proceedings of the Working Conference on Advanced Visual Interface—AVI '02, 123.* http://doi.org/10.1145/1556262.1556281.

Carver, C. A., Howard, R. A., & Lane, W. D. (1999). Addressing different learning styles through course hypermedia. *IEEE Transactions on Education, 42*(1), 33–38.

Ding, Y., Toma, I., Fried, M., Kang, S.-J., & Yan, Z. (2008). Integrating social tagging data: Upper Tag Ontology (UTO). *IEEE International Conference on Systems, Man and Cybernetics, 2008. SMC 2008,* 460–466.

Dunn, R., Dunn, K., & Freeley, M. E. (1984). Practical applications of the research: Responding to students' learning styles–step one. *Illinois State Research and Development Journal, 21*(1), 1–21.

Felder, R., & Silverman, L. (1988). Learning and teaching styles in engineering education. *Engineering Education, 78,* 674–681. http://doi.org/10.1109/FIE.2008.4720326.

Felder, R., Silverman, L., & Solomon, B. (2000). Index of Learning Styles (ILS). *Skynet.ie.*

Graf, S. (2007). *Adaptivity in learning management systems focussing on learning styles.* Vienna University of Technology.

Herlocker, J. L., Konstan, J. A., Borchers, A., & Riedl, J. (1999). An algorithmic framework for performing collaborative filtering. In *Proceedings of the 22nd Annual International ACM SIGIR Conference on Research and Development in Information Retrieval* (pp. 230–237).

Herlocker, J. L., Konstan, J. A., Terveen, L. G., & Riedl, J. T. (2004). Evaluating collaborative filtering recommender systems. *ACM Transactions on Information Systems.* http://doi.org/10.1145/963770.963772.

Ivanović, M., Pribela, I., Vesin, B., & Budimac, Z. (2008). Multifunctional environment for E-learning purposes. *Novi Sad Journal of Mathematics, 38*(2), 153–170.

Janssen, J., den Berg, B., Tattersall, C., Hummel, H., & Koper, R. (2007). Navigational support in lifelong learning: Enhancing effectiveness through indirect social navigation. *Interactive Learning Environments, 15*(2), 127–136.

Kinshuk, K., Chang, M., Dron, J., Graf, S., Kumar, V., Lin, O., Yang, G. et al. (2011). Transition from e-learning to u-learning: innovations and personalization issues. In *Technology for Education (T4E), 2011 IEEE International Conference on* (pp. 26–31).

Klašnja-Milićević, A., Vesin, B., Ivanovic, M., & Budimac, Z. (2011a). Integration of recommendations and adaptive hypermedia into java tutoring system. *Computer Science and Information Systems, 8*(1), 211–224. http://doi.org/10.2298/CSIS090608021K.

Klašnja-Milićević, A., Vesin, B., Ivanović, M., & Budimac, Z. (2011b). E-learning personalization based on hybrid recommendation strategy and learning style identification. *Computers & Education, 56*(3), 885–899.

Kolb, D. (1984). *Individuality in learning and the concept of learning styles* (pp. 61–98). Englewood Cliffs, New Jersey: Prentice Hall.

Kuljis, J., & Liu, F. (2005). A comparison of learning style theories on the suitability for e-learning. *Web Technologies, Applications, and Services, 2005*, 191–197.

Manouselis, N., Drachsler, H., Vuorikari, R., Hummel, H., & Koper, R. (2011). Recommender systems in technology enhanced learning. In *Recommender Systems Handbook* (pp. 387–415). http://doi.org/10.1007/978-0-387-85820-3.

Milicevic, A. K., Nanopoulos, A., & Ivanovic, M. (2010). Social tagging in recommender systems: A survey of the state-of-the-art and possible extensions. *Artificial Intelligence Review, 33*, 187–209. http://doi.org/10.1007/s10462-009-9153-2.

Northrup, P. (2001). A framework for designing interactivity into web-based instruction. *Educational Technology, 41*(2), 31–39.

Pashler, H., McDaniel, M., Rohrer, D., & Bjork, R. (2009). Learning styles concepts and evidence. *Psychological Science in the Public Interest, Supplement, 9*, 105–119. http://doi.org/10.1111/j.1539-6053.2009.01038.x.

Pi-lian, W. T. H. E. (2005). Web log mining by an improved aprioriall algorithm. *Engineering and Technology, 4*(2005), 97–100.

Pritchard, A. (2013). *Ways of learning: Learning theories and learning styles in the classroom.* Routledge.

Tang, T. Y., & McCalla, G. (2005). Smart recommendation for an evolving e-learning system: Architecture and experiment. *International Journal on Elearning, 4*(1), 105.

Vesin, B., Ivanovic, M., Klašnja-Milićević, A., & Budimac, Z. (2011). Rule-based reasoning for altering pattern navigation in programming tutoring system. In *System Theory, Control, and Computing (ICSTCC), 2011 15th International Conference on* (pp. 1–6).

Vesin, B., Ivanović, M., Klašnja-Milićević, A., & Budimac, Z. (2012). Protus 2.0: Ontology-based semantic recommendation in programming tutoring system. *Expert Systems with Applications, 39*, 12229–12246. http://doi.org/10.1016/j.eswa.2012.04.052.

Zervas, P., & Sampson, D. G. (2013). The effect of users' tagging motivation on the enlargement of digital educational resources metadata. *Computers in Human Behavior.* http://doi.org/10.1016/j.chb.2013.06.026.

Part V
Evaluation and Discussion

Evaluation and Incentives

Chapter 12
Experimental Evaluation of Protus 2.1

Abstract Implemented Protus 2.1 for Java programming language has been used in real-life educational environments. The experiments were realized on an educational dataset, consisting of 440 learners, 3rd year undergraduate students of the Department of Information technology at Higher School of Professional Business Studies, University of Novi Sad. The experiment lasted for two semesters. Involved learners were programming beginners that successfully passed the basic computer literacy course at previous semester. They were divided into two groups: the experimental group and the control group. Learners of the control group learned with the previous version of the system and did not receive any recommendation or guidance through the course, while the learners of the experimental group were required to use Protus 2.1 system. Learners from both groups did not take any parallel traditional course and they were required not to use any additional material or help. This chapter highlights the results of the evaluation and discussion of analysis of the results regarding the validity of the tutoring system presented in the previous chapters.

Implemented Protus 2.1 for Java programming language has been used in real-life educational environments. The experiments were realized on an educational dataset, consists of 440 learners, 3rd year undergraduate students of the Department of Information technology at Higher School of Professional Business Studies, University of Novi Sad. The experiment lasted for two semesters. Involved learners were programming beginners that successfully passed the basic computer literacy course at previous semester. They were divided into two groups: the experimental group and the control group. Learners of the control group learned with the previous version of the system (Vesin et al. 2009) and did not receive any recommendation or guidance through the course, while the learners of the experimental group were required to use Protus 2.1 system. Learners from both groups did not take any parallel traditional course and they were required not to use any additional material or help.

© Springer International Publishing Switzerland 2017
A. Klašnja-Milićević et al., *E-Learning Systems*,
Intelligent Systems Reference Library 112, DOI 10.1007/978-3-319-41163-7_12

Table 12.1 The analysis of the test score difference

Type of test	Group	N	Df	Mean	t-calculated value	t-table value
Intellectual abilities	Experimental	340	438	117,25	1.23	1.96
	Control	100	–	111,69	–	–

Level of significance $\alpha = 0.05$

12.1 Data Set for Experiment

The experimental group consisted of 340 learners, while the control group consisted of 100 learners.

In order to assess whether the means of two groups are statistically different from each other, the t-test was utilized. Both groups of learners completed the Norm-referenced test which allows us to compare learners' intellectual abilities (Glaser 1963). Results of this test were combined with grades that learners earned at a basic computer literacy course at the first semester of their studies. The aim of a computer literacy course was to teach data structures and algorithms by presenting exercises of algorithm simulations to the learners.

For these learners programming coursework in any programming language was not assessed. The most important outcome was therefore the introduction of general problem solving concepts, rather than focusing on teaching the syntax of a specific programming language. The predetermined alpha level adopted for hypothesis testing was 0.05, as significance levels of less than 0.05 are considered statically significant, degrees of freedom (df) for the test was 438. Table 12.1 reports the obtained t-test results. Since the calculated value of t (1.23) is not greater than table value of t (1.96), we can conclude that the differences between the experimental and the control group are negligible and there is no need for additional equalization of groups.

12.2 Data Clustering

The learners from the experimental group filled out the Felder-Soloman Index of Learning Styles Questionnaire (Felder 2005) (Fig. 12.1). The aim was to cluster learners from the experimental group into a sub-class according to the learning style. Figure 12.2 shows the comparison of learners' stated preferences corresponding to learning styles across all four domains.

Based on the results of the questionnaires it was possible to define appropriate clusters, which determined learner profiles for 340 learners from the experimental group. Clusters were formed for different combinations of learning styles within the three categories (Table 12.2). Category *Information processing* was omitted in order to increase the number of learners in a separate cluster and to obtain more relevant data for recommendations. In future research, when increasing the number

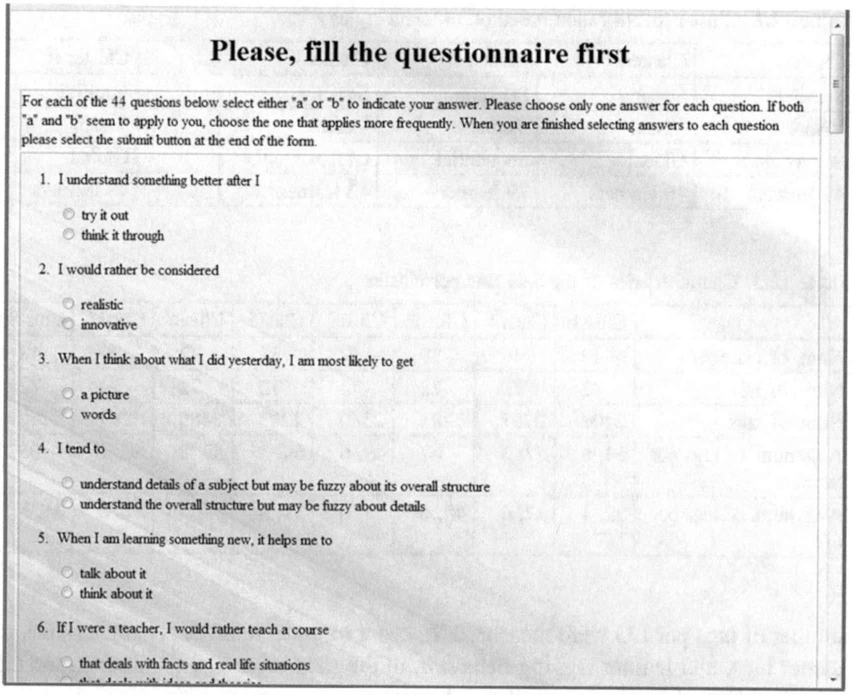

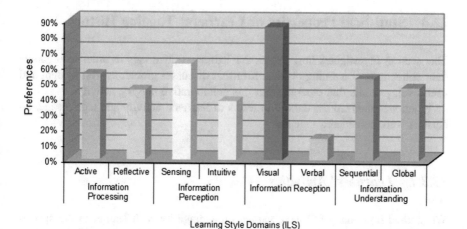

Please, fill the questionnaire first

For each of the 44 questions below select either "a" or "b" to indicate your answer. Please choose only one answer for each question. If both "a" and "b" seem to apply to you, choose the one that applies more frequently. When you are finished selecting answers to each question please select the submit button at the end of the form.

1. I understand something better after I
   - try it out
   - think it through

2. I would rather be considered
   - realistic
   - innovative

3. When I think about what I did yesterday, I am most likely to get
   - a picture
   - words

4. I tend to
   - understand details of a subject but may be fuzzy about its overall structure
   - understand the overall structure but may be fuzzy about details

5. When I am learning something new, it helps me to
   - talk about it
   - think about it

6. If I were a teacher, I would rather teach a course
   - that deals with facts and real life situations
   - that deals with ideas and theories

Fig. 12.1 ILS questionnaire

Fig. 12.2 Learning styles results

of learners participating in the experiment it can be taken into account. The summary use of data per each cluster is shown on Table 12.3. Number of learners, number of LOs, number of tags, average number of tags per learners and average

Table 12.2 Cluster identification based on different styles

Cluster1	Cluster2	Cluster 3	Cluster 4	...	Cluster 8
Sensing	Sensing	Sensing	Sensing		Intuitive
Visual	Visual	Verbal	Verbal		Visual
Sequential	Global	Sequential	Global		Global
49 learners	46 learners	39 learners	42 learners	...	48 learners

Table 12.3 Characteristics of the data sets per cluster

	Clust1	Clust2	Clust3	Clust4	Clust5	Clust6	Clust7	Clust8
Num. of learners	49	46	39	42	35	42	39	48
Num. of LO	72	72	72	72	72	72	72	72
Num. of tags	2402	2707	3283	2380	2243	2486	2289	2268
Avg. num. of tags per learners	54, 6	57, 3	67	49, 6	64, 1	59, 2	58, 7	63
Avg. num. of tags per LO	33, 4	37, 6	45, 6	33, 6	31, 6	34, 5	31, 8	31, 5

number of tags per LO were measured. In order to understand the characteristics of learner tags, and learner tagging behavior, in the next section, we will examine tag characteristics of learners in Protus 2.1 system.

12.3   Statistical Properties of Learners' Tagging History

This section investigates how Protus 2.1 learners utilize tags in order to organize their collections of learning objects. It further discusses global, as well as item and user-level, patterns that emerge from this collaborative tagging activity.

When we analyzed the dataset in terms of learners' activity and tags' usage, all clusters were considered together.

12.3.1 Learners' Activities

We studied how many LO were tagged on average by each learner in the system (Fig. 12.3) and found that 12 % of the learners tagged less than 10 LOs (low activity) 23 % tagged between 10 and 50 LOs (medium activity) and 65 % tagged between 50 and 72 LOs (high activity). We also analyzed the tagging vocabulary, i.e., how many different tags each learner used to define her/his preferred LOs (Fig. 12.4). We found that 21 % of the learners used less than 20 different tags, 71 % used between 20 and 35 tags and the remaining 8 % used between 35 and 65 tags.

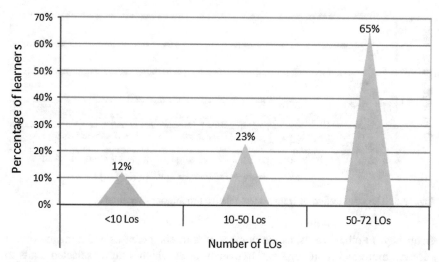

Fig. 12.3 Learner activities on LOs

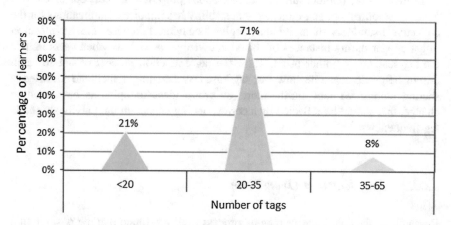

Fig. 12.4 Learners activities on tags

12.3.2 Tag Usage

In order to understand the characteristics of learner tags, and learner tagging behaviour, we examined tag characteristics of learners in Protus 2.1 system. Figure 12.5 illustrates the frequency of use of the first fifteen popular tags in the learner tag set. The X-axis denotes the number of learners entered over time. The Y-axis denotes the proportion of tag usage in a specific time. Here, each line represents a tag.

Overall, these tag patterns reveal that after examining about 20 learners, the frequency of learner tags tends to remain stable at fixed proportions of the total tag

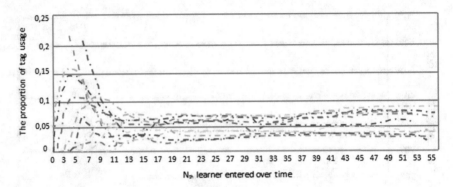

Fig. 12.5 The stabilization of learner tags' relative proportions over time

frequency. In other words, as the number of learners increases and a tagged object receives more and more tags, the frequency of at which a tag is selected tends to become fixed (Golder and Huberman 2006).

In their work, (Golder and Huberman 2006) proposed the concept of convergence or stabilization, and indicated that stability has important implications for the collective usefulness of individual tagging behavior. Likewise, this stabilization might appear during instances of shared knowledge, as well as when users imitate the tag selection of other users. Thus, the tag extraction process of our research design helps demonstrate how our tag-based system might facilitate knowledge sharing among learners. Furthermore, as group ideas or opinions on a reading change, this should be reflected by a corresponding change in the previously stable tag frequencies.

12.3.3 Tag Entropy Over Time

Research on the collaborative tagging process itself has found that the position of a tag correlates with its expressiveness. Tag entropy is a measurement of specificity where more general tags should have higher entropies because they might appear in different topics, whereas seldom tags are often more specific to a topic, thus have lower entropies. The entropy of a tag is defined as:

$$H(T) = -\sum_{t \in T} p(t) \log_2 p(t)$$

Here, T is the set of tags in the profile of the learner, p(t) is the probability that the tag t was utilized by learner and $\log_2 p(t)$ is called self-information. Using base 2 for the computation of the logarithm allows to measure self-information as well as entropy in bits.

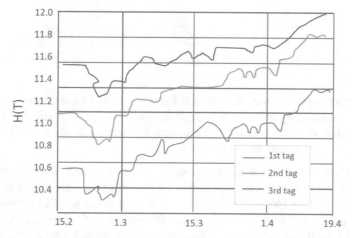

Fig. 12.6 Tag entropy H(T) over time

Figure 12.6 shows the strong correlation between the position and the informativeness of a tag.

It appears that learners tend to assign common tags at the beginning of the tagging process and more specific tags later. There exist, at least three potential explanations for this effect (Wetzker et al. 2010):

1. The affinity to label from general to specific could be a universal behavioral pattern of humans that exists in other domains.
2. The effect could also be a consequence from users' intention to classify new content into a set of relatively constant categories. Adding frequent category tags at the beginning and content specific tags later would result in an increase in entropy as the one observed in Fig. 12.6.
3. Finally, the perceived association between tag position and entropy could be initiated by the tag recommendation functionality.

12.3.4 Semantic Analysis of Tags

A semantic analysis of tags was performed to better understand different utilization of tags. Tags were classified according to (Sen et al. 2006) that is also based on the categories of (Golder and Huberman 2006):

1. **Factual tags**—tags may be used to identify the topic of an object using nouns and proper nouns (e.g. operators, loop, arrays) or to classify the type of object (e.g. tutorial, task, example, basic info, explanation, definition),
2. **Subjective tags**—tags may be used to denote the qualities and characteristics of the item (e.g. useful, interesting, difficult, easy, understandable, blurry) and

Table 12.4 Types analysis of each tag

• PERSONAL	• SUBJECTIVE	• FACTUAL
• 44 %	• 40 %	• 16 %

3. **Personal tags**—item ownership, self-reference, tasks organization—a subset of tags often used by individuals to organize their own learning objects. Much like self-referencing tags, some tags are used by individuals for task organization (e.g. to read, to practise, to print).

When we analyzed how these tags were used and re-used among learners, we found the vast majority of the tags (Table 12.4) were of the personal (44 % of tags) and subjective type (40 % of tags). The rest of the tags (16 % of tags) were factual in their nature and could be used to identify the topic of a learning object. The obtained distribution indicates the fact that learners adapt learning objects themselves and organize them for easily managing. In a MovieLens study (Sen et al. 2006), for comparison, the distribution was 63 % factual 29 % subjective, 3 % personal and 5 % other.

12.4 Experimental Protocol and Evaluation Metrics

The performance of the proposed models is evaluated by holding out a part of the data set as ground-truth data (the test set), and building prediction models from the remaining data (the training set).

We randomly divided the data set into a training set and a test set with sizes 80 and 20 % of the original set, respectively. As performance measures for the item and tag recommendations, we use the classic metrics of precision and recall which are standard in such scenarios (Herlocker et al. 2004). Precision and recall have been in use to evaluate information retrieval systems for many years. Mapping in a recommender system manner, precision and recall have the following definitions regarding the evaluation of top-N recommendations.

For a test user that receives a list of N recommended tags (top-N list), precision and recall are defined as follows:

- **Precision** is the ratio of the number of relevant tags in the top-N list (i.e., those in the top-N list that belong in the future set of tags posted by the test user) to N.
- **Recall** is the ratio of the number of relevant tags in the top-N list of the total number of relevant tags (all tags in the future set posted by the test user).

With i being the item from the randomly picked post of user u and $\widehat{T}(u, i)$ the set of recommended tags, recall and precision can be calculated as:

$$recall\left(\widehat{T}(u,\,i)\right) = \frac{1}{|U|}\sum_{u\in U}\frac{\left|tags(u,\,i)\cap\widehat{T}(u,\,i)\right|}{|tags(u,\,i)|}$$

$$precision\left(\widehat{T}(u,\,i)\right) = \frac{1}{|U|}\sum_{u\in U}\frac{\left|tags(u,\,i)\cap\widehat{T}(u,\,i)\right|}{|(u,\,i)|}$$

All experiments are repeated 10 times and we report the mean of the runs. For each run, we use exactly the same train/test splits.

12.5 Evaluation of Several Suitable Recommendation Techniques

The classical measures precision and recall were chosen to evaluate the performance of several suitable recommendation techniques for RS in e-learning environments.

First, we describe the specific settings used to run evaluated algorithms. Then we present and discuss the results of Protus 2.1 evaluation.

12.5.1 Settings of the Algorithms

Before starting full experimental evaluation of selected algorithms we determined the sensitivity of appropriate parameters to different algorithms and from the sensitivity plots we fixed the optimum values of these parameters and used them for the rest of the experiments. The analysis was performed for all eight clusters. Given the similarity of the obtained values of parameters the results of the first cluster are shown only.

Most Popular Tags. We counted how many post's tags occur globally and used the top tags as recommendations.

Most Popular Tags by Item. For a given item we counted for all tags in how many posts they occur together with that item. We then used the tags that occurred most often together with that item as a recommendation.

Most Popular Tags ρ—Mix. Before comparing *Most Popular Tags ρ—Mix* algorithm with the others, we focused on finding an appropriate size of the parameter ρ. Hence, we observed a similar precision/recall behavior for all values of $\rho \in \{0, 0.1, \ldots, 0.9, 1\}$. As can be seen in Fig. 12.7, variation of algorithm with the most popular tags by user ($\rho = 0$) performs worse than a variety of algorithm with the most popular tags by item ($\rho = 1$) for all numbers of recommended tags. All mixed versions perform better than most popular tags by user and all mixed

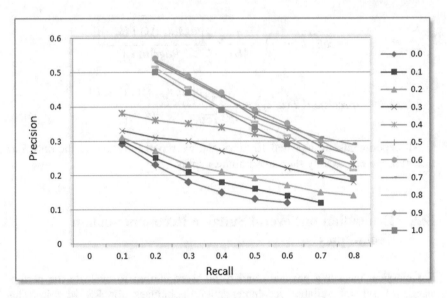

Fig. 12.7 Precision and recall of most popular tags ρ mix for $\rho \in \{0, 0.1, \ldots, 0.9, 1\}$

versions with $\rho \geq 0.5$ perform better than most popular tags by item. The best performance is obtained if $\rho = 0.6$.

Adapted PageRank. With the parameter d = 0.7 we stopped computation after 10 iterations or when the distance between two consecutive weight vectors was less than 10^{-6}. In $\vec{p}$, we gave higher weights to the user and the item from the post which was chosen. While each user, tag and item got a preference weight of 1, the user and item from that particular post got a preference weight of $1 + |U|$ and $1 + |I|$, respectively.

FolkRank. The same parameter and preference weights were used as in the *Adapted PageRank.*

Collaborative Filtering (CF) based on Tags. For *Collaborative Filtering* algorithm the neighbourhood is computed based on the user-tag matrix $\pi_{UT}Y$. The only parameter to be tuned in the CF based algorithms is the number k of best neighbors (Sarwar et al. 2001). We examine the effect of the variation of recalls according to the neighbourhood size k which is closely connected with tag preference generation. Figure 12.8 shows a graph of how recall changes as the neighbour size grows from 10 to 90. Recommender quality initially improves as we increase the neighbourhood size from 10 to 30. However, after the neighbourhood of size 30, increasing the value of k did not lead to statistically significant improvements. That is, once the number of nearest neighbours, k, is sufficiently large, the recommendation quality for each user is not changed by any further increases in the number of nearest neighbours. Considering this trend, we selected 30 as our optimal choice of the neighbourhood size.

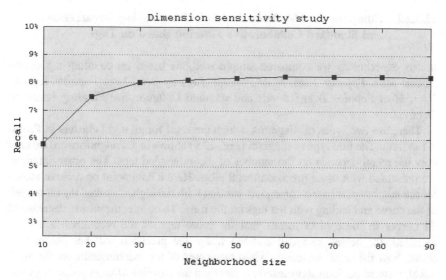

Fig. 12.8 Recall of collaborative tag-based CF according to the variation of neighborhood size

HOSVD—based model. Since there is no straightforward way to find the optimal values for *c1*, *c2* and *c3*, we follow the way according to (Symeonidis et al. 2010) that a 70 % of the original diagonal of $X(1)$, $X(2)$ and $X(3)$ matrices can give good approximations. Thus, *c1*, *c2* and *c3* are set to be the numbers of singular values by preserving 70 % of the original diagonal of $X(1)$, $X(2)$ and $X(3)$ respectively in each run.

RTF. We ran *RTF* with $(k_u, k_i, k_t) \in \{(8, 8, 8); (16, 16, 16); (32, 32, 32)\}$ dimensions, as in (Rendle et al., 2009). The corresponding model is called "*RTF* 8", "*RTF* 16", and "*RTF* 32". The other hyper parameters are: learning rate $\alpha = 0.5$, regularization $\gamma = \gamma_c = 10^{-5}$, iterations *iter* = 500. The model parameters $\hat{\theta}$ are initialized with small random values drawn from the normal distribution N (0, 0.1).

12.5.2   Results of Selected Methods Evaluation

In the following subsection, we will present and discuss the results of selected methods evaluation. First, we compare simple methods based on counting tag occurrences (*Most Popular Tags*), specific approaches for improving the performance of such methods and an adaptation of *User-based Collaborative Filtering*, named *Collaborative Filtering based on Tags*. Then, we analyse the prediction quality of graph-based approaches, *Adapted PageRank* and *FolkRank*, and tensor based approaches, *HOSVD* and *RTF*. Finally, we give a comparative analysis of the best representation of these considered techniques, together.

12.5.2.1 Comparison of Methods Based on Counting Tag Occurrences and Standard Collaborative Filtering Based on Tags

In our experiments we compared simple methods based on counting tag occurrences: *Most Popular Tags*, *Most Popular Tags by Item*, *Most Popular Tags by User*, *Most Popular Tags 0.6–mix* and standard *Collaborative Filtering based on Tags*.

There are two types of diagrams, which are used for all eight clusters in Protus 2.1 system. The first type of diagram (Fig. 12.9) shows in a straightforward manner how the recall depends on the number of recommended tags. The other diagrams are presented with usual precision/recall plots. Here a data point on a curve stands for the number of tags recommended, starting with the highest ranked tag on the left of the curve and ending with ten tags on the right. Therefore, the steady decrease of all curves in those plots means that the more tags of the recommendation are regarded, the better the recall and the worse the precision will be. Figure 12.9 shows how the recall increases, when more tags of the recommendation are used. All algorithms perform significantly better than the baseline *Most Popular Tags* and the *Most Popular Tags by User* strategy, whereas it is much harder to beat the *Most Popular Tags by Item*. The idea to suggest the most popular tags by item results in a recall which is very similar to using the CF recommender based on user's item similarities. In contrast to these two approaches, the *Most Popular Tags ρ Mix-*

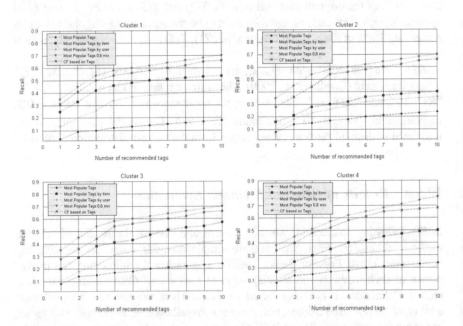

Fig. 12.9 Recall for methods based on counting tag occurrences as a function of number of recommended tags for the eight clusters of Protus 2.1 system

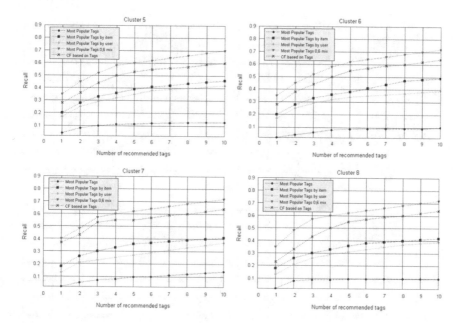

Fig. 12.9 (continued)

recommender includes also the user's tags in the recommendations. As the diagrams show, it is successful and could gain results better than those of CF. The precision-recall plots in Fig. 12.10 extend diagrams from Fig. 12.9 with the precision measure. All algorithms perform significantly better than the baseline *Most Popular Tags* and the *Most Popular Tags by User* strategy. It is remarkable that the *Most Popular Tags 0.6–Mix* recommender provides on average better precision and recall than both *Collaborative Filtering* algorithms and *Most Popular Tags by Item*.

12.5.2.2   Comparison of Graph-Based Approaches

We saw in Sect. 7.4 that in order to apply standard CF-based algorithms to folksonomies, some data transformation must be performed. Such transformations lead to information loss, which can lower the recommendation quality. Another well-known problem with CF-based methods is that large projection matrices must be kept in memory, which can be time/space consuming and thus compromise real time recommendations. Also, for each different mode to be recommended, the algorithm must be eventually changed, demanding an additional effort for offering multi-mode recommendations.

FolkRank builds on *PageRank* and proved to give significantly better tag recommendations than CF, because *FolkRank* has ability to exploit the information

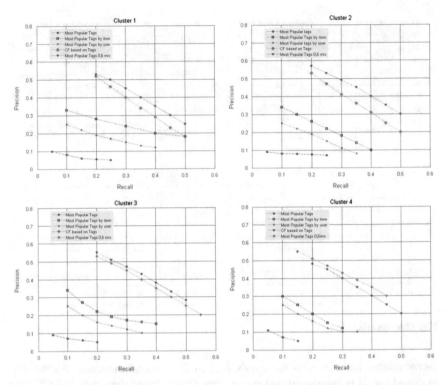

Fig. 12.10 Recall and precision for methods based on counting tag occurrences for eight clusters of Protus 2.1 system

that is appropriate to the specific user together with input from other users via the integrating structure of the underlying hypergraph. When comparing the prediction quality of CF, Adapted PageRank and *FolkRank* (Fig. 12.11) one can see that *FolkRank* outperform both two. *FolkRank* is able to predict, additionally to globally relevant tags, the exact tags of the user which CF could not. This is due to the fact that *FolkRank* considers, via the hypergraph structure, also the vocabulary of the user himself, which CF by definition doesn't do. This method also allows for mode switching with no change in the algorithm. Moreover, as well as CF-based algorithms, *FolkRank* is robust against online updates since it does not need to be trained every time a new user, item or tag enters the system. However, *FolkRank* is computationally expensive and not trivially scalable, making it more suitable for systems where real-time recommendations are not a requirement.

12.5.2.3 Comparison of Methods Based on Tensor Factorization

Similarly, tensor factorization methods also work directly over the ternary relation of the folksonomy. Although the tensor reconstruction phase can be expensive, it

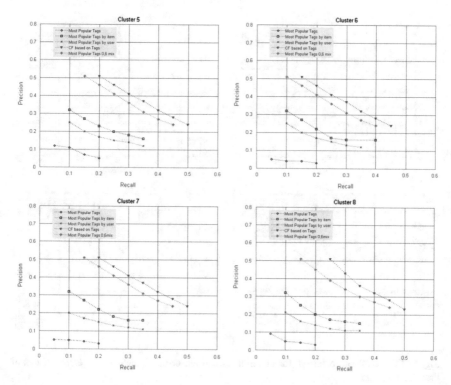

Fig. 12.10 (continued)

can be performed offline. After the lower dimensional tensor is computed, the recommendations can be done fast, making these algorithms suitable for real-time recommendations. A potential disadvantage of tensor factorization methods is that easy mode switching can only be achieved if one considers that the different recommendation problems, i.e., user/item/tag, can be addressed by minimizing the same error function. If one chooses *HOSVD* for example, the reconstructed tensor can be used for multi-mode recommendations with trivial mode switching, but at the cost of eventually solving the wrong problem: *HOSVD* minimizes a least-square error function while social tagging RS are more related to ranking. If one tries to optimally reconstruct the tensor with regard to an error function targeted to a specific recommendation mode on the other hand, accuracy is eventually improved, but at the cost of making mode switching more involved. Even though *RTF* and *HOSVD* have the same prediction method and thus prediction complexity, in practice *RTF* models are much faster in prediction than comparable *HOSVD* models, because *RTF* models need much less dimensions than *HOSVD* for attaining better quality. Also, for the task of personalized ranking *HOSVD* has three major drawbacks to *RTF* (Rendle et al. 2009):

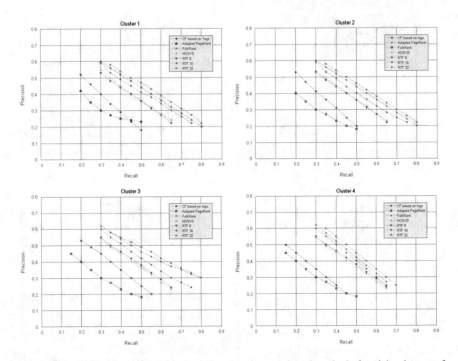

Fig. 12.11 Recall and precision for graph-based and tensor-based methods for eight clusters of Protus 2.1 system

1. *HOSVD doesn't take into account missing values.* For tag recommendation the missing values are usually filled with zeros (Symeonidis et al. 2008).
2. *HOSVD optimizes for minimal element-wise error.* But for the ranking problem of tag recommendation, we are interested in another objective function.
3. *HOSVD has no regularization.* For machine learning tasks, preventing over-fitting is very important so *HOSVD* is predisposed to overfitting.

A final problem with *HOSVD* is sensitivity to the number of dimensions and that they have to be chosen carefully. Also *HOSVD* is sensitive to the relations between the user, item and tag dimensions (e.g. choosing the same dimension for all three dimensions leads to poor results). In contrast to this, for *RTF* it can be chosen the same number of dimensions for user, item and tag. Besides this theoretical analysis, in Fig. 12.11 it can be seen that the prediction quality of *RTF* is clearly better to the one of *HOSVD*. Also, Fig. 12.11 shows that even with a very small number of 8 dimensions, *RTF* achieves almost similar results as *HOSVD*. Increasing the dimensions of *RTF* to 16 dimensions, it already outperforms *HOSVD* in quality. Furthermore, for *RTF*, by increasing the number of dimensions we get better results. When comparing the prediction quality of *RTF* and *FolkRank* (Fig. 12.11) one can

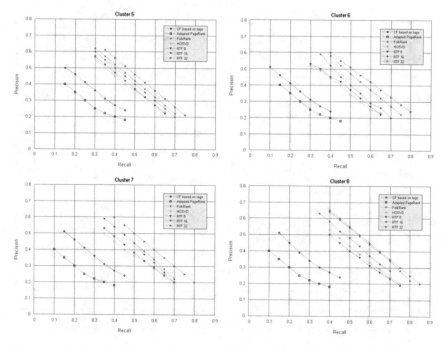

Fig. 12.11 (continued)

see that *RTF* with 8/16 dimensions achieves comparable results whereas 32 dimensions outperform FolkRank in quality.

12.5.2.4 Summary of Algorithms' Advantages and Disadvantages

According to the conducted experiments on real-life dataset and educational environment, in this section, we briefly discuss the main advantages and disadvantages of the aforementioned algorithms. Standard CF-based algorithms need some data transformation in order to apply to folksonomies. Such transformations lead to information loss, which can lower the recommendation quality. Another problem with CF-based methods is that large projection matrices must be kept in memory, which can be time and space overwhelming and thus compromise real-time recommendations. Also, for each different mode to be recommended, the algorithm must be eventually improved, demanding an additional effort for offering multi-mode recommendations. As well as CF-based algorithms, *FolkRank* is robust against online updates since it does not need to be trained every time a new user, item or tag enters the system. However, *FolkRank* is computationally expensive and not trivially scalable, making it more suitable for systems where real-time

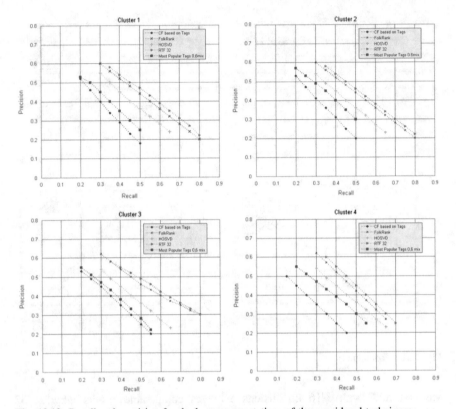

Fig. 12.12 Recall and precision for the best representatives of the considered techniques

recommendations are not a requirement. *FolkRank* also allows mode switching with no change in the algorithm. Similarly, tensor factorization methods work directly over the ternary relation of the folksonomy. Although the tensor reconstruction phase can be expensive it can be performed offline. After the lower dimensional tensor is computed, the recommendations can be done quickly, making these algorithms appropriate for real-time recommendations. A possible drawback of tensor factorization methods is that easy mode switching can only be achieved if one considers that the different recommendation problems, i.e., user/item/tag, can be addressed by minimizing the same error function. If one chooses *HOSVD* for example, the reconstructed tensor can be used for multi-mode recommendations with simple mode switching, but at the cost of solving the wrong problem: *HOSVD* minimizes a least-square error function while social tagging RS is more related to ranking. Figure 12.12 shows a comparison between some of the aforementioned algorithms. We selected only the best representatives of the considered techniques. We can conclude that the best method is *RTF* followed by *FolkRank* and *HOSVD*.

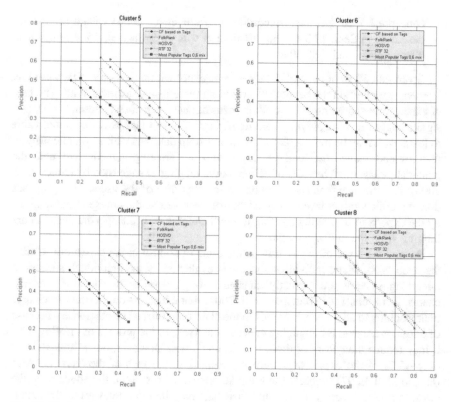

Fig. 12.12   (continued)

12.6   Expert Validity Survey

In order to systematically verify the relationship between learning comprehension and learner data tagging, and in order to help teachers evaluate learner knowledge, an experimental test was created with the expert tag set.

The collection of learner tags was compared with the tags given independently by 4 experts in the field. The expert tag set was comprised of 165 tags of which 100 were different.

Within expert tag set, we elaborated two research questions:

1. Which learning objects can be found by a simulated query with the expert tags on the complete set of learner tags and which relevance (number of matching tags) does it have? With respect to this question, we found that the ratio of matches was in average 45 % of the expert tags also assigned to a learning object by the learners.
2. How many keywords assigned as tags are already present as text in the LO? This question addresses if the tags given to the learning items stay close to the original item. The results were that experts tend to tag more abstractly and

conceptual then learners. According to (Sen et al. 2006) categories (as we described in Sect. 12.3.4), the distribution was 73 % factual, 16 % subjective, 4 % personal and 7 % other.

Given that 55 % of the expert tags were tags not within the body of tags used by learners, we question the benefits of providing these tags to learners at all. The lack of expert time and willingness to fill in metadata has been cited (Friesen 2001), as a significant hurdle to deploying learning objects. If expert tags provide limited value to learners, it may be more appropriate to bootstrap data sets with automatic tagging features and reduce the load on those who are creating content. We note the potential pedagogical benefits of collaborative tagging as suggested by (Jones et al. 2006): that the tags themselves represent the expertise of the users. This proposes that at a collaborative level, a tag set can be observed as the course is being given by the experts to improving an insight into the topics and concepts that learners are filtering from the online material.

Beyond the issue of expert time is the issue of control in the classroom. Unlike the open Web, where individual success is evaluated by the individual, success in e-learning systems are typically dictated through a series of educator prepared exams. It has observed (Bateman et al. 2007) that educators are hesitant to change their teaching to adopt new methods in the classroom (virtual or otherwise), because of a loss of control. By engaging educators actively in the process of creating tags, it may reduce their fears of these new technologies. However, our results showed only 45 % of the expert tags were represented in the tags of the learners. Also (Halpin et al. 2007) suggested that unlike open a Web system, the educator in the classroom is not merely a peer, and their tags may be more relevant to the examinations, which may be useful to learners.

12.7 Evaluation of Protus 2.1 System from the Educational Point of View

Educational research measures are needed to evaluate whether learners actually do benefit from the usage of the recommender system. From the educational point of view, learners only benefit from learning technology when it makes learning more effective, efficient or attractive. Efficiency indicates the time that learners needed to reach their learning goal. Effectiveness is a measure of the total amount of completed, visited, or studied lessons during a learning phase (Drachsler et al. 2009). In our study, we track only lessons that are successfully completed, meaning that learners passed the appropriate test at the end of the particular lesson. It is related to the efficiency variable through counting the actual study time. To answer this question, we randomly selected a sample of 100 learners from the experimental group and 100 learners from the control group. The results of the experiment showed that the learners in the experimental group should be able to complete a course in less time than learners in the control group who learned with the previous

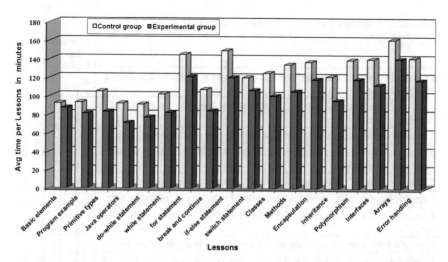

Fig. 12.13 Efficiency comparison between groups

version of the system (Fig. 12.13). Figure 12.14 shows that the experimental group continuously completed more lessons successfully than the control group.

Satisfaction reflects the individual satisfaction of learners with the given recommendations. Satisfaction is closely related to the motivation of the learner and therefore a rather important measure of learning. To get a subjective evaluation of our system, at the end of the course we organized a non-mandatory questionnaire that collected learners' (from the experimental group) opinions about the main features of the system. The results of the questionnaire were used to improve the quality of lessons. This questionnaire (Table 12.5) maps a set of 16 questions over 4 dimensions: 'ease of tagging', 'usefulness of tagging', 'usefulness of Protus 2.1 system and tag exchange', and 'Protus 2.1 system is easy-to-use'.

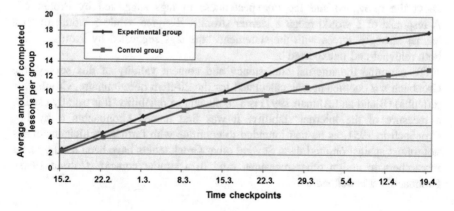

Fig. 12.14 Average completions of lessons per group

Table 12.5 Questionnaire—Analysis of satisfaction with Protus 2.1 system

Question	Response				
	1	2	3	4	5
I can easily construct meaningful words or phrases to represent the learning objects with tags					
I can clearly indicate the meaning of tags which describes the context of learning objects					
The proposed tags allow me to express the right term when entering tags					
Reviews of my previous own tags facilitate the recall of learning objects' ideas and information					
The tags suggested by Protus 2.1 are both relevant and unexpected					
I am comfortable with others knowing what I mean about learning object					
I am interested in seeing others who are likeminded to me regarding specific topics					
Having a high level of influence on my neighbours is important to me					
I think the using tags enable me to easily grasp the structure and concepts of learning objects					
I think that exploiting tagging information provide me with effective feedback during the learning process					
The tagging activities inspire me to make new ideas					
I think that Protus 2.1's user interface is simple to learn (understand) and efficient to use					
I can start quickly with Protus 2.1 system					
I believe that Protus 2.1 is effective at learning my preferences					
I am satisfied with Protus 2.1's recommendations					
I trust that Protus 2.1 has ability to make correct recommendations for me					

Out of 100 learners, 75 filled in the questionnaire. Participants are asked to give a level of agreement to each question on a 1–5 scale (strongly disagree, disagree, neutral, agree, and strongly agree). For example, the fifth question in Table 12.5 is about the relevance and the unexpectedness of tags suggested by Protus 2.1. A response of 1 would mean a learner strongly disagree, while 5 would mean a learner strongly agrees with the statement: "the tags suggested by Protus 2.1 are both relevant and unexpected".

To examine the internal consistency and content validity of this survey, the Cronbach's α coefficient was calculated for the 20-item questionnaire. Cronbach's α (alpha) (Bland and Altman 1997) is a coefficient of reliability. It is usually used as a measure of the internal stability. It was first named Cronbach's α by Lee Cronbach in 1951, as he had intended to continue with further coefficients. α a is not robust against omitted data. Several other Greek letters have been used by later researchers to assign other measures used in a similar context (Cortina 1993). Cronbach's α is defined as:

Table 12.6 Rotated factor loadings and Cronbach's α value for four factors

Items	Factor1	Factor2	Factor3	Factor4
Factor 1: Easy to tagging $\alpha = 0.776$				
I_1	0.672	–	–	–
I_2	0.727	–	–	–
I_3	0.812	–	–	–
Factor 1: Usefulness of tagging $\alpha = 0.833$				
I_6	–	0.614	–	–
I_7	–	0.718	–	–
I_8	–	0.588	–	–
Factor 1: Usefulness of tagging $\alpha = 0.716$				
I_9	–	–	0.731	–
I_{10}	–	–	0.708	–
I_{11}	–	–	0.645	–
I_{12}	–	–	0.811	–
Factor 1: Usefulness of tagging $\alpha = 0.768$				
I_{15}	–	–	–	0.871
I_{16}	–	–	–	0.738
I_{17}	–	–	–	0.645
I_{18}	–	–	–	0.821
I_{19}	–	–	–	0.672

$\alpha = 0.814$, total variance explained is 68.12 %

$$\alpha = \frac{K}{K-1}\left(1 - \frac{\sum_{i=1}^{K}\sigma_{Y_i}^2}{\sigma_X^2}\right)$$

where K is the number of *items* or *testlets*, σ_X^2 is the variance of the observed entire test scores, and $\sigma_{Y_i}^2$ the variance of item i for the current sample of persons (DeVellis 1991). In order to determine if a question item is correlated with a factor, we applied the distinguish validity test by using the factor analysis method to observe each question item. Four factors among these items are shown in Table 12.6. The eigenvalues of the four factors are greater than 1.00 with variance 68.12 % explained. From the experimental results, it was found that some question items, were not correlated with factors (that is, their load was less than 0.5). As a result, 4 questions items were dropped, reducing the overall number to 16. In addition, the experiment shows that the internal reliability indexes of the four factors are 0.776, 0.833, 0.716, and 0.768, respectively.

Table 12.7 Statistical results of the questionnaire for evaluating Protus 2.1 system

Questionnaire item (four factors)	Strongly disagree (%)	Disagree (%)	Neutral (%)	Agree (%)	Strongly agree (%)
Easy to tagging in Protus 2.1	–	16.81	42.19	38.31	2.69
Usefulness of tagging in Protus 2.1	–	4.35	27.19	53.14	15.32
Usefulness of Protus 2.1 system	–	–	25.71	68.1	6.19
Protus 2.1 is easy-to-use	–	–	46	47.8	6.1

The α coefficient is 0.814 after deleting non-correlated factors. Therefore, these results suggest that these factors were sufficiently reliable for representing learner tagging behaviours, when the Cronbach's α is higher than 0.7 (Hwang et al. 2008; Nunnally et al. 1967).

The statistical analysis of the survey results is summarized in Table 12.7. The major findings are presented as follows:

1. 97 % of the learners indicated that creating tags were easy, and that it was easy to construct meaningful words or phrases to represent the learning objects with tagging objects.
2. 85 % of the learners thought that tagging activity can help learners summarize new ideas and quickly grasp the structure and concepts. Some learners indicated that their tags were more accurate after sufficient tagging practice.
3. 93 % of the learners agreed that Protus 2.1 is capable of helping them to easily comprehend the context of learning objects, and can help them improve their learning efficiency.
4. 94 % of the learners regarded that Protus 2.1 system is easy-to-use.

References

Bateman, S., Brooks, C., McCalla, G., & Brusilovsky, P. (2007). Applying collaborative tagging to e-learning. *WWW*, 1–7. http://doi.org/10.1.1.64.8892.

Bland, J. M., & Altman, D. G. (1997). Statistics notes: Cronbach's alpha. *Bmj, 314*(7080), 572.

Cortina, J. M. (1993). What is coefficient alpha? An examination of theory and applications. *Journal of Applied Psychology*. http://doi.org/10.1037/0021-9010.78.1.98.

DeVellis, R. F. (1991). Guidelines in scale development. In *Scale Development* (pp. 51–60).

Drachsler, H., Hummel, H. G. K., & Koper, R. (2009). Identifying the goal, user model and conditions of recommender systems for formal and informal learning. *Journal of Digital Information, 10*(2).

Felder, R. M. (2005). A study of the reliability and validity of the Felder-Soloman index of learning styles. *Engineering Education, 113*, 77. http://doi.org/10.1109/IIAI-AAI.2015.284.

Friesen, N. (2001). What are educational objects? *Interactive Learning Environments, 9*(3), 219–230.

Glaser, R. (1963). Instructional technology and the measurement of learing outcomes: Some questions. *American Psychologist, 18*(8), 519.

Golder, S. A., & Huberman, B. A. (2006). The structure of collaborative tagging systems. *Journal of Information Science, 32*(2), 198–208.

Halpin, H., Robu, V., & Shepherd, H. (2007). The complex dynamics of collaborative tagging. In *Proceedings of the 16th International Conference on World Wide Web* (pp. 211–220).

Herlocker, J. L., Konstan, J. A., Terveen, L. G., & Riedl, J. T. (2004). Evaluating collaborative filtering recommender systems. In: *ACM Transactions on Information Systems*. http://doi.org/10.1145/963770.963772.

Hwang, G.-J., Tsai, P.-S., Tsai, C.-C., & Tseng, J. C. R. (2008). A novel approach for assisting teachers in analyzing student web-searching behaviors. *Computers & Education, 51*(2), 926–938.

Jones, N., Macasek, M., Walonoski, J., Rasmussen, K., & Heffernan, N. (2006). Common tutor object platform–an e-learning software development strategy. In *Proceedings of the 15th International Conference on World Wide Web, Edinburgh* (pp. 307–316). Scotland.

Nunnally, J. C., Bernstein, I. H., & Berge, J. M. F. T. (1967). *Psychometric theory* (Vol. 226). JSTOR.

Rendle, S., Balby Marinho, L., Nanopoulos, A., & Schmidt-Thieme, L. (2009). Learning optimal ranking with tensor factorization for tag recommendation. In *Proceedings of the 15th ACM SIGKDD International Conference on Knowledge Discovery and Data Mining* (pp. 727–736).

Sarwar, B., Karypis, G., Konstan, J., & Riedl, J. (2001). Item-based collaborative filtering recommendation algorithms. In *Proceedings of the 10th International Conference on World Wide Web* (Vol. 1, pp. 285–295). http://doi.org/10.1145/371920.372071.

Sen, S., Lam, S. K., Rashid, A. M., Cosley, D., Frankowski, D., Osterhouse, J.,& Riedl, J. (2006). Tagging, communities, vocabulary, evolution. In *Proceedings of the 2006 20th Anniversary Conference on Computer Supported Cooperative Work* (pp. 181–190).

Symeonidis, P., Nanopoulos, A., & Manolopoulos, Y. (2010). A unified framework for providing recommendations in social tagging systems based on ternary semantic analysis. *Knowledge and Data Engineering, IEEE Transactions on, 22*(2), 179–192.

Symeonidis, P., Ruxanda, M. M., Nanopoulos, A., & Manolopoulos, Y. (2008). Ternary semantic analysis of social tags for personalized music recommendation. In *ISMIR* (Vol. 8, pp. 219–224).

Vesin, B., Ivanović, M., & Budimac, Z. (2009). Learning management system for programming in java. *Annales Universitatis Scientiarum De Rolando Eötvös Nominatae, Sectio-Computatorica, 31*, 75–92.

Wetzker, R., Zimmermann, C., Bauckhage, C., & Albayrak, S. (2010). I tag, you tag: Translating tags for advanced user models. In *Proceedings of the Third ACM International Conference on Web Search and Data Mining* (pp. 71–80).

Chapter 13
Conclusions and Future Directions

Abstract E-learning is an important segment of educational environments. It represents a unique opportunity to learn independently, regardless of time and place, to acquire knowledge without interruption and customized to the individual and based on the principles of traditional education. Today, the most popular forms of e-learning are: web-based e-learning systems, virtual classrooms or tutoring systems. This monograph presents how the Semantic web technologies, ontologies and adaptation rules can be used to improve the performance of an existing tutoring system. The architecture of a personalized tutoring system that relies entirely on Semantic Web technologies and standards is presented. Ontologies that correspond to the components of the traditional tutoring system are shown in detail. This chapter concludes the monograph, summarizing the main contributions and discussing the possibilities for future work.

E-learning is an important segment of educational environments. It represents a unique opportunity to learn independently, regardless of time and place, to acquire knowledge without interruption and customized to the individual and based on the principles of traditional education. Today, the most popular forms of e-learning are: Web-based e-learning systems, Virtual classrooms or Tutoring systems.

E-learning systems use various techniques to generate recommendations for selection of appropriate teaching materials and activities to learners based on their needs, skills and learning styles. Recommendation systems in e-learning adapts educational materials and/or user interface to the specific needs and demands of learners.

Recommender systems made significant progress over the last decade when numerous content-based, collaborative, and hybrid methods were proposed and several "industrial-strength" systems have been developed. However, despite all of these advances, the current generation of recommender systems still requires further improvements to make recommendation methods more effectively in a broader range of applications. With the increasing popularity of the collaborative tagging systems, surveyed in this monograph, tags could be interesting and useful information to enhance recommender systems' algorithms. Besides helping users

© Springer International Publishing Switzerland 2017
A. Klašnja-Milićević et al., *E-Learning Systems*,
Intelligent Systems Reference Library 112, DOI 10.1007/978-3-319-41163-7_13

organize his or her personal collections, a tag also can be regarded as a user's expression, while tagging can be considered as an implicit rating or voting on the tagged information or items. Thus, the tagging information can be used to make recommendations.

Monograph presents approaches for the development of a general model of the personalized Web tutoring system for attending courses in various domains. On the basis of this model Protus 2.1 system was developed. This system automatically adapts instructional materials and user interface to the requirements, habits and knowledge level of each learner. Differences between learners are determined based on the current level of knowledge, individual learning styles, characteristics, requirements and goals of learners. The system automatically directs the activities of learners and generates referral links, actions and teaching materials in the learning process.

The main research directions of the monograph can therefore be summarized as follows.

Chapter 1 started with an introduction of the monograph research aim. The chapter further highlighted research objectives of the monograph.

Chapter 2 presents the basis of electronic learning techniques for personalization of the learning process and the possibilities of their integration in e-learning systems.

Chapter 3 presents the bases of electronic learning techniques for personalization of learning process based on individual learning styles and the possibilities of their integration in e-learning systems.

Chapter 4 shows the most popular forms of adaptation of educational materials to learners. Details for personalisation based on recommender systems, link adaptation and learning style identification are presented.

Chapter 5 presents current trends for use of intelligent agents for personalization in e-learning systems.

Chapter 6 provided a comprehensive survey of the state-of-the-art in recommender systems, collaborative tagging systems and folksonomy for tagging activities which can be used for extending the capabilities of recommender systems.

Chapter 7 presents a theoretical overview of tag-based recommender systems in e-learning environments and identifies the limitations of the current generation of collaborative tagging techniques and discusses about some approaches for extending their capabilities.

Chapter 8 contains a review of the basic elements of Semantic Web, as well as the possibilities for application of Semantic Web technologies in e-learning.

Chapter 9 displays the details of a general tutoring system model, supported by Semantic Web technologies and the principles of creating courses from different domains supported by this model.

Chapter 10 contains the details about previous versions of the system, defined user requirements for the new version of the system, architectural details, and general principles for application of defined general tutoring model for implementation of programming courses in Protus 2.1.

Chapter 11 considers adaptation based on learning styles and possibilities for applying a recommender system based on collaborative tagging techniques in developing a Protus 2.1 tutoring system that adapts to the interests and level of learners' knowledge in the various fields.

Chapter 12 analyses the statistical properties of learners' tagging history and presents an evaluation of the performance of several suitable recommendation techniques for RS in e-learning environments and comparison of describing techniques. Also, this chapter considers expert tag set in order to systematically verify the relationship between learning comprehension and learner data tagging, and in order to help teachers evaluate learner knowledge. Finally, in this chapter evaluation of Protus 2.1 system from the educational point of view is performed.

Chapter 13 concludes the monograph, summarizing the main contributions, and discussing the possibilities for future work.

13.1   Contributions of the Monograph

This monograph presents how the Semantic Web technologies, ontologies and adaptation rules can be used to improve the performance of an existing tutoring system. The architecture of a personalized tutoring system that relies entirely on Semantic Web technologies and standards is presented. Ontologies that correspond to the components of the traditional tutoring system are shown in detail.

Presented system's architecture fully supports the use of Semantic Web technologies for building basic elements of the system and defining the personalization rules.

Ontologies will completely change the way systems are designed and organized. A large knowledge base is still formed without possibilities of information sharing and reuse. In the future, the development of intelligent tutoring systems will facilitate an extensive library of ontologies. Instead of development of such systems from scratch, it will be possible to use system components extracted from existing libraries and repositories. This process will shorten development time and improve the robustness and reliability of newly created knowledge base and tutoring systems itself.

The explicit conceptualization of system components in the form of ontologies, promote the exchange and reuse of knowledge, and communication and cooperation between system components. Improved the use of ontology to construct systems that require explicitly structured knowledge. Such systems allow learners to access larger collection of information and resources. Displayed architecture is modular and allows greater flexibility and the possibility for replacement of individual components until they correspond to the current interface. Defined ontologies can serve as a foundation of knowledge that can be further extended and modified in order to define adaptive systems from different domains.

Although ontologies have a set of basic implicit reasoning mechanisms derived from the description logic, which they are typically based on (such as classification, relations, instance checking, etc.), they need rules to make further inferences and to express relations that cannot be represented by ontological reasoning. Thus, ontologies require a rule system to derive/use further information that cannot be captured by them, and rule systems require ontologies in order to have a shared definition of the concepts and relations mentioned in the rules. The rules also allow adding expressiveness to the representation formalism, reasoning on the instances, and they can be orthogonal to the description logic on which ontologies are based on.

As a part of Web 2.0, collaborative tagging is getting popular as an important tool to classify dynamic content for searching and sharing. We analyzed the potential of collaborative tagging systems, including personalized and biased user preference analysis, and specific and dynamic classification of content for applying collaborative tagging techniques into Java Tutoring system. Appropriate selection of collaborative tagging techniques could lead to applying the best results in terms of increasing motivation in learning process and understanding of the learning content. The scientific contributions are summarized as follows. First, in this monograph, we demonstrated how programming tutoring systems can be enabled to provide adaptivity based on learning styles. We introduced a general concept for a tutoring system to automatically generate course that fit to the learning styles of the learners. The only additional effort from the teachers and course developers is to provide some meta-data in order to annotate the learning material. Furthermore, learners were asked to fill out the ILS questionnaire for detecting their learning styles. The concept was implemented and an experiment with 440 learners was performed to show the effectiveness of the realized concept. Then, we evaluated statistical properties of learners' tagging history. We studied how many LO were tagged on average by each learner in the system and found that even 65 % learners show high activity, tagged between 50 and 72 LOs. In order to understand the characteristics of learner tags, and learner tagging behaviour, we examined tag characteristics of learners in Protus 2.1 system. We have noted: as the number of learners increases and a tagged object receives more and more tags, the frequency of at which a tag is selected tends to become fixed. This concept of convergence or stabilization has important implications for the collective usefulness of individual tagging behaviour. Likewise, this stabilization might appear during instances of shared knowledge, as well as when learners imitate the tag selection of other learners. Research on the collaborative tagging process itself has found that the position of a tag correlates with its expressiveness. Tag entropy is a measurement of specificity where more general tags should have higher entropies because they might appear in different topics, whereas seldom tags are often more specific to a topic, thus have lower entropies. It appears that learners tend to assign common tags at the beginning of the tagging process and more specific tags later. A semantic analysis of the tags was performed to better understand different utilization of tags. When we analyzed how these tags were used and re-used among learners, we found the vast majority of the tags were of the personal (44 % of the tags) and subjective type (40 % of the tags).

The most significant part of the research focuses on appropriate selection of collaborative tagging techniques which could lead to applying the best results in terms of increasing motivation in learning process and understanding of the learning content. As a result, personalized and the most likely preferred recommendations can be estimated to an active learner that are in accordance with the learner's interests, his learning style, demographic characteristics and previously acquired knowledge. First, we compare simple methods based on counting tag occurrences (Most Popular Tags), specific approaches for improving the performance of such methods and an adaptation of the *User-based Collaborative Filtering*, named *Collaborative Filtering based on Tags*. All algorithms perform significantly better than the baseline *Most Popular Tags* and the *Most Popular Tags by User* strategy, whereas it is much harder to beat the *Most Popular Tags by Item*. The idea to suggest the *Most Popular Tags by Item* results in a recall which is very similar to using the CF recommender based on user's item similarities. In contrast to these two approaches, the *Most Popular Tags 0.6* Mix-recommender includes also the learner's tags in the recommendations. It is successful and could gain results better than those of CF.

Then, we analysed the prediction quality of graph-based approaches, Adapted PageRank and *FolkRank*, and tensor based approaches, *HOSVD* and *RTF*. *FolkRank* builds on PageRank and proved to give significantly better tag recommendations than CF, because *FolkRank* has ability to exploit the information that is appropriate to the specific learner together with input from other learners via the integrating structure of the underlying hypergraph. When comparing the prediction quality of CF, *Adapted PageRank* and *FolkRank* one can see that *FolkRank* outperform both too. This method also allows for mode switching with no change in the algorithm. Moreover, as well as CF-based algorithms, *FolkRank* is robust against online updates since it does not need to be trained every time a new learner, item or tag enters the system. However, *FolkRank* is computationally expensive and not trivially scalable, making it more suitable for systems where real-time recommendations are not a requirement. Similarly, tensor factorization methods also work directly over the ternary relation of the folksonomy. Although the tensor reconstruction phase can be expensive, it can be performed offline. After the lower dimensional tensor is computed, the recommendations can be done quickly, making these algorithms appropriate for real-time recommendations. A possible drawback of tensor factorization methods is that easy mode switching can only be achieved if one considers that the different recommendation problems, i.e., learner/item/tag, can be addressed by minimizing the same error function. If one chooses *HOSVD* for example, the reconstructed tensor can be used for multi-mode recommendations with simple mode switching, but at the cost of solving the wrong problem: *HOSVD* minimizes a least-square error function while social tagging RS is more related to ranking. We selected only the best representatives of the considered techniques. We concluded that the best method is *RTF* followed by *FolkRank* and *HOSVD*.

Also, we have carried out other experiments to evaluate the performance of the system from the points of view of both teachers and learners. The results demonstrated the potential pedagogical benefits of collaborative tagging that the tags themselves represent the expertise of the users. This proposes that at a collaborative level, a tag set can be observed as the course is being given by the experts to improve an insight into the topics and concepts that learners are filtering from the online material. The general opinion of experts has been very positive. They have demonstrated a high degree of motivation and have especially liked the novelty of using learners' data to improve e-learning courses, to be able to apply modifications to courses directly from the system and have the possibility of working and sharing information with other teachers and educational experts. However, experts have indicated that the creation of the repository or knowledge database is a hard task.

From the educational point of view, learners only benefit from learning technology when it makes learning more effective, efficient or attractive. The results of the experiment showed that the learners who were required to use Protus 2.1 system should be able to complete a course in less time than learners in the control group who learned with the previous version of the system. Also, these learners continuously completed more lessons successfully than the control group. To get a subjective evaluation of our system, at the end of the course we organized a non-mandatory questionnaire that collected learners' opinions about the main features of the system. The results are very encouraging:

1. Learners indicated that creating tags were easy, and that it was easy to construct meaningful words or phrases to represent the learning objects with tagging objects.
2. Learners thought that tagging activity can help learners summarize new ideas and quickly grasp the structure and concepts. Some learners indicated that their tags were more accurate after sufficient tagging practice.
3. Learners agreed that Protus 2.1 is capable of helping them to easily comprehend the context of learning objects, and can help them improve their learning efficiency.
4. Learners regarded that Protus 2.1 system is easy-to-use.

The main achievement of this work is the following:

- defining the general tutoring system model using Semantic Web technologies,
- design of separate components of personalized tutoring system in the form of educational ontologies,
- explicit display of adaptation rules for easier understanding, update and reuse of tutoring system components,
- implementation of various personalization options in a tutoring system, and
- presented possibilities for use of Semantic Web technologies to build a tutoring system.

This architecture provides a foundation for further expansion of the system and allows detailed modelling of the personalization process. Also, the adaptation rules can be modified in order to achieve specific requirements during the personalization process and learner modelling.

Presented tutoring system architecture enables the implementation of other programming courses or courses from other domains with minimal changes to the defined architecture. Changes would include the addition of new educational materials and modification options for testing learners' knowledge.

13.2 Future Work and Open Research Questions

The rapid development of collaborative tagging system and related emerging technology suggests new ideas for personalized recommendation and determine a great number of challenges for future work.

Future studies could focus more specifically on measuring the impact of prior learner experience (with computers and the Internet) and interest (in the knowledge domain) on the effect of creating tags. Additionally, future studies could investigate whether there are more factors which also have an influence on the effect of choice of tags. Possible candidates could be a mood or stress level.

Improvements in the experimental design could verify the findings reported in this monograph and increase their external validity. Similar comparative studies could be carried out involving more learners and more experts and teachers from other areas (unrelated to computer science) in order to obtain a more heterogeneous teacher's profile. This will allow the study of other interesting questions such as: Is it possible that different teachers in different areas might coincide in their evaluation of patterns?; What is the behaviour of experts and teachers as they progress through a course?; Can tuples that are found to be valid and useful in one course later be applied to another course with a different profile? These aspects could lead to a confirmation that would focus uniquely on a detailed analysis of the changes made and whether the process is efficient and likely to be corresponding to non-guided course content revision.

Even though the source code was written specifically for the Java programming course that was used in the experimental evaluation, it is feasible that with adequate programming effort, adapted versions of Protus 2.1 can be created for other knowledge domains. We can, also, integrate other sequence mining algorithms (Han et al. 2005) such as SPADE, FreeSpan, CloSpan and PSP, and other clustering algorithm without demanding the learner to specify any parameter. We plan to evaluate the quality of the recommendations based on feedback from learners as well as on results using a testing set of data. Finally, it would be very useful to develop a real-time feedback loop between data mining and the recommendation system. We can use, for example, intelligent agents for doing on-line data mining automatically and for communicating with the recommender systems. In this way the system could work completely autonomously. The agents can mine data only when they notice enough volume of new data. And the authors do not have to pre-process and apply mining algorithms; they only have to organize the new recommender links if they want.

In the domain of tensor factorization for social tagging Recommender Systems as a recent and prominent field, the research study of this area has just begun to expose the benefits that those methods have to propose. A mainly interesting research direction considers investigating tensor factorization models that highlight both high recommendation accuracy and easy mode switching. As emphasized before, folksonomies usually do not contain numerical ratings, but recently the

GroupLens[1] research group released a folksonomy dataset in which numerical ratings for the tagged items are also given (Marinho et al. 2011). This represents several research opportunities on how to exploit the item's rating information in order to improve recommendations. In this case, a single data structure for all the modes, such as tensors or hyper-graphs, would eventually fail since the ratings are only related to user-item pairs and not to tags. Similar issues can be investigated for content-based methods. It is proven that content-based methods usually neglect the user information, but past research shows that hybrid methods that combine user preferences with the item's content usually conduct to better recommenders. Here, again, tensor or hyper-graph representations would be unsuccessful since items' content is only related to the items but not to the users or tags. So hybrid-based methods that achieve some kind of synthesis between folksonomy representations and items' content would be appreciated contribution to the area.

Finally, Protus 2.1 proposed a new, dynamic approach to adaptive behavior in learning style-responsive environments. Future work will deal with an in-depth analysis of the results with respect to different learning style dimensions as well as the different adaptation features. We also plan to add more adaptation features to our concept and implement them. Another future direction will be to combine the proposed concept with an automatic learner modelling approach so that the system is able to automatically detect the learning styles of the learners based on their behaviour and actions in the system.

References

Han, J., Pei, J., & Yan, X. (2005). Sequential pattern mining by pattern-growth: Principles and extensions*. In *Foundations and Advances in Data Mining* (pp. 183–220). Springer.

Marinho, L. B., Nanopoulos, A., Schmidt-Thieme, L., Jäschke, R., Hotho, A., Stumme, G., Symeonidis, P. et al. (2011). Social tagging recommender systems. In *Recommender systems handbook* (pp. 615–644). Springer.

---

[1]http://www.grouplens.org/.

Printed in the United States
By Bookmasters